Methods and Applications of Error-Free Computation

R. T. Gregory
E. V. Krishnamurthy

Springer-Verlag
New York Berlin Heidelberg Tokyo

R. T. Gregory
Department of Computer Science
 and Department of Mathematics
University of Tennessee
Knoxville, TN 37996
U.S.A.

E. V. Krishnamurthy
Department of Applied Mathematics
Indian Institute of Science
Bangalore 560012
India

Series Editor

David Gries
Department of Computer Science
Cornell University
Upson Hall
Ithaca, NY 14853
U.S.A.

With 1 Figure

(C.R.) Computer Classification: F.2

Library of Congress Cataloging in Publication Data
Gregory, R. T. (Robert Todd)
 Methods and applications of error-free computation.
 (Texts and monographs in computer science)
 Bibliography: p.
 Includes index.
 1. Approximation theory—Data processing. 2. Floating-point arithmetic.
I. Krishnamurthy, E. V. II. Title. III. Series.
QA297.5.G735   1984   511′.4′0285   84-1281

Typeset by Asco Trade Typesetting Ltd., Hong Kong.
Printed and bound by R. R. Donnelley & Sons, Harrisonburg, Virginia.
Printed in the United States of America.

9 8 7 6 5 4 3 2 1

ISBN 0-387-90967-2 Springer-Verlag New York Berlin Heidelberg Tokyo
ISBN 3-540-90967-2 Springer-Verlag Berlin Heidelberg New York Tokyo

To Margaret and Bani

Preface

This book is written as an introduction to the theory of error-free computation. In addition, we include several chapters that illustrate how error-free computation can be applied in practice. The book is intended for seniors and first-year graduate students in fields of study involving scientific computation using digital computers, and for researchers (in those same fields) who wish to obtain an introduction to the subject.

We are motivated by the fact that there are large classes of ill-conditioned problems, and there are numerically unstable algorithms, and in either or both of these situations we cannot tolerate rounding errors during the numerical computations involved in obtaining solutions to the problems. Thus, it is important to study finite number systems for digital computers which have the property that computation can be performed free of rounding errors.

In Chapter I we discuss single-modulus and multiple-modulus residue number systems and arithmetic in these systems, where the operands may be either integers or rational numbers. In Chapter II we discuss finite-segment p-adic number systems and their relationship to the p-adic numbers of Hensel [1908]. Each rational number in a certain finite set is assigned a unique Hensel code and arithmetic operations using Hensel codes as operands is mathematically equivalent to those same arithmetic operations using the corresponding rational numbers as operands. Finite-segment p-adic arithmetic shares with residue arithmetic the property that it is free of rounding errors.

In Chapters III, IV, V, and VI we describe how error-free computation can be used to obtain the exact solutions to certain classes of problems in numerical linear algebra.

The explicit prerequisites of this book consist for the most part of an exposure to elementary number theory, elementary numerical algorithms, and a course in linear algebra.

We use Roman numerals to number the chapters and we use decimal notation to number the sections in each chapter as well as the items in each section. Thus, equations, theorems, corollaries, tables, figures, etc., are numbered consecutively as items within sections. For example, the fifteenth item in Section 2 (of any chapter) is numbered 2.15. This item might be referenced in several ways. For instance, if it is an *equation*, it is referenced as (2.15). If it is any other kind of item (such as a *lemma*), it is called Lemma 2.15. If the item is in Section 2, Chapter V, and the reference is made in some other chapter, it is referenced as Lemma 2.15, Chapter V. Within Chapter V it is referenced as Lemma 2.15, with no further designation.

Finally, we wish to acknowledge with appreciation the contributions of those who in various ways influenced the book: Jo Ann Howell, T. M. Rao, K. Subramanian, Ruth Ann Lewis, Shu-Hwa Hwang, and John Smyre, who were our students; Peter Kornerup, who proved Theorem 6.39, Chapter I, and gave us an efficient way to carry out the inverse mapping, and who provided the first author with excellent research facilities and support during his two visits to Aarhus; Germund Dahlquist, who provided the first author with excellent research facilities and support during his visit to Stockholm; Azriel Rozenfeld, who provided the second author with encouragement and support during his many visits to the University of Maryland; David Matula and Carl Gregory, who made major contributions to Section 7, Chapter I; Arne Fransén, who read the entire manuscript with great care and made many valuable suggestions for improvement; the University of Tennessee, Knoxville, the Curriculum Development Cell (Ministry of Education and Culture, India) and the Indian Institute of Science, Bangalore, for financial assistance towards the preparation of the manuscript; Suzie Witenbarger, for her excellent job of typing the entire manuscript; and our wives, to whom this book is dedicated.

August 1983   R. T. Gregory
                                                              E. V. Krishnamurthy

Contents

Glossary of Notation

A^T	the transpose of A
A^H	the complex conjugate transpose of A
A^{-1}	the inverse of A
A^-	a g-inverse of A
A_R^-	a reflexive g-inverse of A
A_L^-	a least-squares g-inverse of A
A_M^-	a minimum-norm g-inverse of A
A^+	the Moore–Penrose g-inverse of A
A_I^-	A_R^-, if $AA_R^- \in \mathbb{I}^{mm}$ and $A_R^- A \in \mathbb{I}^{nn}$
$\|A\|_m$	if $A = (a_{ij})$, $\|A\|_m = (\|a_{ij}\|_m)$
$\|a\|_m$	the least non-negative residue of a modulo m
$/a/_m$	the symmetric residue of a modulo m
$\|a/b\|_m$	$\|a \cdot b^{-1}(m)\|_m$
$\|a\|_\beta$	$[\|a\|_{m_1}, \|a\|_{m_2}, \ldots, \|a\|_{m_n}]$
$/a/_\beta$	$[/a/_{m_1}, /a/_{m_2}, \ldots, /a/_{m_n}]$
$\left\|\dfrac{a}{b}\right\|_\beta$	$\left[\left\|\dfrac{a}{b}\right\|_{m_1}^*, \left\|\dfrac{a}{b}\right\|_{m_2}^*, \ldots, \left\|\dfrac{a}{b}\right\|_{m_n}^*\right]$
$a^{-1}(m)$ or a^{-1}	the multiplicative inverse of a modulo m
$a^{-1}(\beta)$	$[a^{-1}(m_1), a^{-1}(m_2), \ldots, a^{-1}(m_n)]$
$\langle a \rangle_\rho$	$\langle d_0, d_1, \ldots, d_{n-1} \rangle$, mixed-radix digits for a
β	$[m_1, m_2, \ldots, m_n]$, base vector
ρ	$[r_1, r_2, \ldots, r_n]$, mixed radices
$\mathbb{F}_N$	the set of order-N Farey fractions
$\gcd(a, b)$	greatest common divisor of a and b
$[a/b]$	the integer part of the quotient a/b

$\|\alpha\|_p$	the p-adic norm of α
$H(p, r, \alpha)$	the ordinary Hensel code for α
$\hat{H}(p, r, \alpha)$	the floating-point Hensel code for α
$\mathbb{I}$	the set of integers
$\mathbb{I}_m$	$\{0, 1, 2, \ldots, m-1\}$
$\hat{\mathbb{I}}_m$	$\{|a/b|_m : a/b \in \mathbb{F}_N\}$
$\mathbb{I}_\beta$	$\{|s|_\beta : s \in \mathbb{I}\}$
$\mathbb{I}^m$	the m-dimensional vectors over $\mathbb{I}$
$\mathbb{I}^{mn}$	the $m \times n$ dimensional matrices over $\mathbb{I}$
$\mathbb{I}_r^{mn}$	the $m \times n$ dimensional matrices over $\mathbb{I}$ of rank r
$\mathbb{Q}$	the set of rational numbers
$\mathbb{Q}_p$	the set of p-adic numbers
$\mathbb{R}$	the set of real numbers
$\mathbb{S}_m$	$\{-(m-1)/2, \ldots, -2, -1, 0, 1, 2, \ldots, (m-1)/2\}$
$\mathbb{S}_\beta$	$\{/s/_m : s \in \mathbb{I}\}$
$\mathrm{tr}(A)$	the trace of A
$\tilde{\mathbb{T}}_\beta$	$\{|a/b|_\beta : a/b \in \mathbb{F}_N\}$

CHAPTER I
Residue or Modular Arithmetic

1 Introduction

Since an automatic digital computer is a finite machine, it is capable of representing, internally, only a finite set of numbers. Thus, any attempt to use an automatic digital computer to do arithmetic in the field of real numbers $(\mathbb{R}, +, \cdot)$ is doomed to failure because $\mathbb{R}$ is an infinite set and most of the elements in this set cannot be represented in a computer.

This does not mean, however, that we do not attempt to *approximate* the arithmetic of $(\mathbb{R}, +, \cdot)$ on a computer. We often attempt this approximation by using the so-called floating-point numbers, $\mathbb{F}$, more appropriately called the set of *computer-representable numbers*. The set $\mathbb{F}$ is a set of real numbers with the following properties:

(a) $\mathbb{F}$ is a finite subset of the rational numbers $\mathbb{Q}$.
(b) $\mathbb{F}$ is usually symmetric with respect to the origin, in which case there are two computer representations of zero.
(c) The elements of $\mathbb{F}$ are not evenly distributed along the real line. The interval between two "adjacent" computer-representable numbers is quite small near the origin and becomes increasingly large as one moves away from the origin. In the vicinity of the largest possible computer-representable number the interval is quite large. For example, see Forsythe, Malcolm, and Moler [1977], p. 12, and Gregory [1980], p. 9.
(d) Almost all of the "familiar" rational numbers are excluded from $\mathbb{F}$. For example, on a binary computer the only candidates for membership in $\mathbb{F}$ are rational numbers of the form p/q, where q is a power of two. Thus, such numbers as $\frac{1}{10}, \frac{1}{3}, \frac{5}{6}$ and $\frac{2}{7}$ are excluded.

1

(e) The system $(\mathbb{F}, +, \cdot)$ does not constitute a field (primarily because we do not have closure under either of the two binary operations indicated).

Thus, there is no possibility of representing the continuum of real numbers in any detail. A "practical" solution to this problem in many cases, is to represent a real number x by its closest computer-representable number $\hat{x}$, thereby introducing the *rounding error*

(1.1) $\varepsilon = x - \hat{x}.$

Because of the lack of closure, rounding errors are also introduced as a result of arithmetic operations on elements of $\mathbb{F}$. For example, if $\hat{x}$ and $\hat{y}$ are two "adjacent" elements in $\mathbb{F}$, then

(1.2) $$z = \frac{\hat{x} + \hat{y}}{2}$$

is not an element of $\mathbb{F}$ and will have to be represented by $\hat{z}$, the element in $\mathbb{F}$ "nearest" to z. In this example, $\hat{z}$ could be either $\hat{x}$ or $\hat{y}$.

To some people, the effect of inexact arithmetic (and of rounding errors) may not appear to be too serious. However, it is well known that, from the point of view of *backward error analysis* (see Young and Gregory [1972], p. 10, for example), the computed solution* to a problem can be interpreted as the exact solution to a slightly perturbed problem. Furthermore, there exists a class of problems, called *ill-conditioned problems* for which the (exact) solution is extremely sensitive to "small" perturbations in the data. In solving such problems, therefore, the introduction of rounding errors can be disastrous.

For example, consider the problem of evaluating the determinant of the matrix

(1.3) $$A = \begin{bmatrix} -73 & 78 & 24 \\ 92 & 66 & 25 \\ -80 & 37 & 10 \end{bmatrix}.$$

That this problem[†] is extremely ill-conditioned is shown by the fact that if

(1.4) $$E = \begin{bmatrix} 0 & 0 & 0 \\ 0 & 0 & 0 \\ 0 & 0 & 10^{-2} \end{bmatrix},$$

then

(1.5) $\det(A) = 1,$

---

* In this discussion we are assuming that the computed solution is obtained by inexact arithmetic on elements of $\mathbb{F}$.

† See Gregory and Karney [1978], p. 50.

whereas

$$(1.6) \qquad \det(A + E) = -118.94.$$

In other words, if an accumulation of rounding errors were to correspond to the introduction of the perturbation matrix E, then the computed value of the determinant would be -118.94, whereas the exact value is 1.

Because of the difficulties associated with attempts at approximating arithmetic in the field of real numbers $(\mathbb{R}, +, \cdot)$ by using the finite set $\mathbb{F}$, and because of the effects of this approximation on our attempts at solving ill-conditioned problems, there is a strong motivation for using number systems with which we can do exact arithmetic. A residue number system is an example of such a system.

2 Single-Modulus Residue Arithmetic

It is well known that automatic digital computers can perform certain arithmetic operations exactly if the operands are integers. This motivates us to consider integer arithmetic as a means of avoiding rounding errors in the hope that certain ill-conditioned problems can be solved exactly. If we do integer arithmetic and reduce the results modulo m, we are using an arithmetic system called *single-modulus residue arithmetic*. Here the integer $m > 1$ is called the *modulus* of the arithmetic system.

Theoretical background

Much of the material which follows is elementary number theory and we shall assume that the reader is familiar with these simple ideas. See, for example, Hardy and Wright [1960]. However, we shall quickly review these ideas and, in doing so, we shall give many theorems without proof. Most of these results, with an emphasis on computer applications, appear in Szabó and Tanaka [1967], and in Young and Gregory [1973], Chapter 13. The symbol $\mathbb{I}$ will denote the set of integers throughout this book.

2.1. Definition.* Let $a, b, m \in \mathbb{I}$, with $m > 1$. If m divides $b - a$, we say b *is congruent to a* modulo m and we write

$$b \equiv a \qquad (\text{mod } m).$$

If m does not divide $b - a$, we say b *is not congruent to a* modulo m and we write

$$b \not\equiv a \qquad (\text{mod } m).$$

---

*Negative integers can be used as moduli but to no apparent advantage because $-m$ divides $b - a$ if and only if m divides $b - a$.

Relations such as $a \equiv b \pmod{m}$ are called *congruences* and there are several properties of congruences reviewed in the following theorems (which we state without proof).

2.2. Theorem. *Let $a, b, m \in \mathbb{I}$, with $m > 1$. Then the following three statements are equivalent:*

(i) $a \equiv b \qquad \pmod{m}$

(ii) $b \equiv a \qquad \pmod{m}$

(iii) $a - b \equiv 0 \qquad \pmod{m}.$

2.3. Theorem. *Let $a, b, c, m \in \mathbb{I}$, with $m > 1$. If*

$$a \equiv b \qquad \pmod{m},$$

and

$$b \equiv c \qquad \pmod{m},$$

then

$$a \equiv c \qquad \pmod{m}.$$

2.4. Theorem. *Let $a, b, c, d, x, y, m \in \mathbb{I}$, with $m > 1$. If*

$$a \equiv b \qquad \pmod{m}$$

and

$$c \equiv d \qquad \pmod{m},$$

then

$$ax + cy \equiv bx + dy \qquad \pmod{m}.$$

2.5. Theorem. *Let $a, b, c, d, m \in \mathbb{I}$, with $m > 1$. If*

$$a \equiv b \qquad \pmod{m}$$

and

$$c \equiv d \qquad \pmod{m},$$

then

$$ac \equiv bd \qquad \pmod{m}.$$

In single-modulus residue arithmetic we map each integer $b \in \mathbb{I}$ onto an integer r in the (finite) set

(2.6) $\mathbb{I}_m = \{0, 1, 2, \ldots, m - 1\}.$

We call r the *least non-negative residue* of b modulo m. The mapping is defined as follows.

2.7. Definition. $|\cdot|_m : \mathbb{I} \to \mathbb{I}_m$ is defined by writing

$$|b|_m = r$$

if and only if $0 \le r < m$ and

$$b \equiv r \quad (\mathrm{mod}\ m).$$

Using this mapping it is easy to see that $\mathbb{I}$ is the union of m disjoint subsets $\mathbb{R}_0, \mathbb{R}_1, \ldots, \mathbb{R}_{m-1}$ called *residue classes*, where

$$(2.8) \qquad \mathbb{R}_k = \{b \in \mathbb{I} : |b|_m = k\}.$$

2.9. Example. If $m = 7$, then $58 \in \mathbb{R}_2$ because

$$|58|_7 = 2.$$

We say that 2 is the least non-negative residue of 58 modulo 7, or that 58 has been *reduced* to 2 modulo 7. The next two theorems demonstrate simple properties of this mapping.

2.10. Theorem. *Let* $a, b, m \in \mathbb{I}$, *with* $m > 1$. *Then*

(i) $\qquad\qquad\qquad |a|_m$ *is unique.*

(ii) $\qquad |a|_m = |b|_m$ *if and only if* $a \equiv b \quad (\mathrm{mod}\ m).$

(iii) $\qquad\qquad |km|_m = 0$ *for every* $k \in \mathbb{I}.$

2.11. Theorem. *Let* $a, b, m \in \mathbb{I}$, *with* $m > 1$. *Then*

(i)
$$\begin{aligned}
|a + b|_m &= \big||a|_m + |b|_m\big|_m \\
&= \big||a|_m + b\big|_m \\
&= \big|a + |b|_m\big|_m.
\end{aligned}$$

(ii)
$$\begin{aligned}
|ab|_m &= \big||a|_m |b|_m\big|_m \\
&= \big||a|_m b\big|_m \\
&= \big|a |b|_m\big|_m.
\end{aligned}$$

Notice that Theorem 2.11 shows us how to do addition and multiplication modulo m. It is clear that there is complete freedom of choice as to when to reduce an operand modulo m. For example, there are four ways to add 24 and 38 modulo 3, and these are equivalent.

(i)
$$\begin{aligned}
|24 + 38|_3 &= |62|_3 \\
&= 2.
\end{aligned}$$

(ii)
$$\begin{aligned}
|24 + 38|_3 &= |0 + 2|_3 \\
&= 2.
\end{aligned}$$

(iii)
$$|24 + 38|_3 = |0 + 38|_3$$
$$= 2.$$

(iv)
$$|24 + 38|_3 = |24 + 2|_3$$
$$= 2.$$

Similar expressions can be written to describe multiplication modulo m.

To see how to do subtraction and division modulo m we need to state a theorem about $\mathbb{I}_m$, the set of non-negative residues modulo m, which was defined in (2.6).

2.12. Theorem. *The system* $(\mathbb{I}_m, +, \cdot)$, *where* $+$ *and* $\cdot$ *denote addition modulo* m *and multiplication* modulo* m, *respectively, constitutes a finite commutative ring with identity.*

PROOF. We merely verify that the following properties are valid for every $a, b, c \in \mathbb{I}_m$.

closure	$|a + b|_m \in \mathbb{I}_m$	$|ab|_m \in \mathbb{I}_m$
commutativity	$|a + b|_m = |b + a|_m$	$|ab|_m = |ba|_m$
associativity	$|a + (b + c)|_m = |(a + b) + c|_m$	$|a(bc)|_m = |(ab)c|_m$
unique identities	$|a + 0|_m = |a|_m$	$|a \cdot 1|_m = |a|_m$
unique additive inverse	$|a + \underline{a}|_m = 0$	. . . .
distributivity	$|a(b + c)|_m = |ab + ac|_m$	

where the *additive inverse of a modulo m* is

$$\underline{a} \equiv |-a|_m$$
$$= m - a. \qquad\qquad \square$$

We can define subtraction in the ring $(\mathbb{I}_m, +, \cdot)$ as the addition of the additive inverse modulo m.

2.13. Definition.

$$|a - b|_m \equiv |a + \underline{b}|_m.$$

It is possible to define a *multiplicative inverse modulo m* for certain elements of $(\mathbb{I}_m, +, \cdot)$ and this makes it possible, in case the divisor is one of those elements, to define division as multiplication by the multiplicative inverse modulo m. However, the big question is "when does a multiplicative inverse modulo m exist?" The following theorem will help us answer this question.

---

*We often omit the symbol $\cdot$ and write ab instead of $a \cdot b$.

2.14. **Theorem.** *The finite commutative ring* $(\mathbb{I}_m, +, \cdot)$ *is a finite field if and only if m is a prime.*

PROOF. See, for example, McCoy [1948], p. 22.   □

Consequently, if m is a prime, $(\mathbb{I}_m, +, \cdot)$ is isomorphic to the Galois field $GF(m)$ and every non-zero element in $\mathbb{I}_m$ has a multiplicative inverse modulo m, which is defined as follows:

2.15. **Definition.** If m is a prime, if $b \neq 0$, and if $b \in \mathbb{I}_m$, there exists a unique integer $c \in \mathbb{I}_m$ which satisfies the equation

$$|cb|_m = |bc|_m = 1.$$

We call c the multiplicative inverse of b modulo m and write

$$c = b^{-1}(m),$$

or simply b^{-1} when the modulus is understood.

If m is not a prime, $(\mathbb{I}_m, +, \cdot)$ is not a field and a non-zero element may or may not have a multiplicative inverse. To know when one exists we need the following theorem and corollary. The expression $\gcd(a, b)$ denotes the *greatest common divisor* of a and b. See Definition 6.1 in this chapter.

2.16. **Theorem.** *Let* $b \in \mathbb{I}$. *Then there exists a unique integer* $c \in \mathbb{I}_m$ *which satisfies*

$$|bc|_m = |cb|_m = 1,$$

if and only if $|b|_m \neq 0$ *and* $\gcd(b, m) = 1$.

2.17. **Corollary.** *If* $b \in \mathbb{I}_m$ *is non-zero, then* $b^{-1}(m) \in \mathbb{I}_m$ *exists (and is unique) if and only if b and m are relatively prime.*

2.18. EXAMPLES. If $m = 10$ (a composite),

$$\mathbb{I}_{10} = \{0, 1, 2, 3, 4, 5, 6, 7, 8, 9\}$$

and only 1, 3, 7, and 9 have multiplicative inverses modulo 10. These are 1, 7, 3, and 9, respectively. If $m = 5$ (a prime),

$$\mathbb{I}_5 = \{0, 1, 2, 3, 4\}$$

and all non-zero elements (that is, 1, 2, 3, and 4) have multiplicative inverses modulo 5. These are 1, 3, 2, and 4, respectively. Obviously, $(\mathbb{I}_5, +, \cdot)$ is isomorphic to the Galois field $GF(5)$.

In summary, then, $(\mathbb{I}_m, +, \cdot)$ is always a finite commutative ring and, in particular, if m is a prime, it is also a finite field isomorphic to the Galois field $\mathrm{GF}(m)$. In either case, if b^{-1} exists, we define division modulo m as follows:

2.19. Definition.

$$\left|\frac{a}{b}\right|_m = |ab^{-1}|_m.$$

We should point out that the quotient of two integers in single-modulus residue arithmetic, when it exists, is always an integer, even in those cases where b does not divide a.

2.20. Example.

$$|7/9|_{11} = |7 \cdot 9^{-1}|_{11}$$
$$= |7 \cdot 5|_{11}$$
$$= 2.$$

In this example, it appears that single-modulus residue arithmetic cannot be used to divide 7 by 9. However, it is not correct to say that the result of this computation is meaningless,* as the following example demonstrates.

2.21. Example.

$$|7/9 \cdot 27|_{11} = \left||7/9|_{11} \cdot |27|_{11}\right|_{11}$$
$$= |2 \cdot 5|_{11}$$
$$= 10.$$

Thus, the integer we computed in Example 2.20 can be used as an intermediate result in the calculation in Example 2.21. This illustrates the fact that single-modulus residue arithmetic can be used in carrying out a sequence of arithmetic operations on integers in $\mathbb{I}_m$ even though the sequence involves one or more division operations, as long as m is relatively prime to each integer which appears in a denominator (so that the appropriate multiplicative inverses modulo m exist).

The only difficulty is in interpreting the computed results. If the correct (mathematical) answer is an integer in $\mathbb{I}_m$, then the result obtained using residue arithmetic will agree with the correct answer. If, on the other hand, the correct answer is not an integer in $\mathbb{I}_m$, then the result obtained using residue arithmetic will not agree with the correct answer and some additional information will be needed in order to obtain the correct answer. Observe that, in the example above, $(\frac{7}{9})(27) = 21$ and $21 \equiv 10 \pmod{11}$.

---

*See Section 5 where Example 2.20 has a different interpretation than the one given here.

2.22. Remark. It is obvious that, if feasible, the modulus m should be a prime, because this guarantees that $(\mathbb{I}_m, +, \cdot)$ is a finite field which implies all non-zero elements in $\mathbb{I}_m$ have multiplicative inverses modulo m.

Applications of the theory

We have defined $|b|_m$ in Definition 2.7 to be the least non-negative residue of b modulo m. With this definition, computation in single-modulus residue arithmetic is simple in the sense that no negative integers are involved in the system $(\mathbb{I}_m, +, \cdot)$. However, if we wish to solve problems such as the determinant evaluation problem described in (1.3), where some of the matrix elements are negative integers, we must be able to handle negative integers as well as positive integers.

One way to accomplish this is to introduce the system of *symmetric residues* modulo m. For symmetry with respect to the origin, m must be an *odd* integer. Hence, if we form the set

$$(2.23) \qquad \mathbb{S}_m = \left\{ -\frac{m-1}{2}, \ldots, -2, -1, 0, 1, 2, \ldots, \frac{m-1}{2} \right\},$$

then each integer $b \in \mathbb{I}$ can be mapped onto an integer $s \in \mathbb{S}_m$ by the following mapping.

2.24. Definition. $/\cdot/_m : \mathbb{I} \to \mathbb{S}_m$ is defined by writing

$$/b/_m = s$$

if and only if

$$b \equiv s \qquad (\mathrm{mod}\ m)$$

and

$$-\frac{m}{2} < s < \frac{m}{2}.$$

We call $/b/_m$ the *symmetric residue* of b modulo m.

It is easily verified that $(\mathbb{S}_m, +, \cdot)$ is a finite commutative ring, and in particular, if m is a prime, $(\mathbb{S}_m, +, \cdot)$ is a finite field. It is also easily verified that $(\mathbb{S}_m, +, \cdot)$ is isomorphic to $(\mathbb{I}_m, +, \cdot)$. Consequently, if the data describing a problem consist of integers in $\mathbb{S}_m$, we can map them into $\mathbb{I}_m$, carry out the computations,* and map the results back into $\mathbb{S}_m$. The mapping functions for doing this are

---

*All computations could be done in $(\mathbb{S}_m, +, \cdot)$, of course, but this requires that algebraic signs be monitored.

$$(2.25) \qquad |a|_m = \begin{cases} /a/_m, & \text{if } 0 \le /a/_m < \dfrac{m}{2} \\ /a/_m + m & \text{otherwise} \end{cases}$$

and

$$(2.26) \qquad /a/_m = \begin{cases} |a|_m, & \text{if } 0 \le |a|_m < \dfrac{m}{2} \\ |a|_m - m & \text{otherwise.} \end{cases}$$

Figure 2.27 illustrates the relationship between $\mathbb{S}_m$ and $\mathbb{I}_m$ for the case $m = 11$.

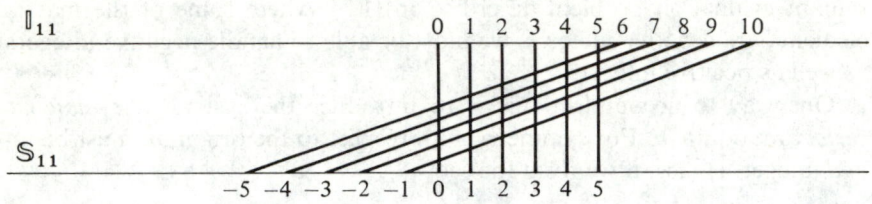

2.27. Figure. The Mappings between $\mathbb{S}_{11}$ and $\mathbb{I}_{11}$.

2.28. **Example.** Let $x = a/b + c$, where $a = 48$, $b = 12$, and $c = -24$. Since we are assuming $a, b, c \in \mathbb{S}_m$ we need to map them into $\mathbb{I}_m$ using (2.25), at the very beginning of the computation. Suppose we let $m = 103$. Then

$$\begin{aligned} |x|_{103} &= |48/12 + (-24)|_{103} \\ &= \left| |48 \cdot 12^{-1}|_{103} + 79 \right|_{103} \\ &= \left| |48 \cdot 43|_{103} + 79 \right|_{103} \\ &= |4 + 79|_{103} \\ &= 83. \end{aligned}$$

Finally, using (2.26), we map this result back into $\mathbb{S}_{103}$ and obtain

$$/x/_{103} = -20,$$

which is the correct answer.

2.29. **Remark.** It is important to choose m large enough so that $\mathbb{S}_m$ contains both the data which describe the problem and the solutions to the problem. Otherwise, some of the solutions may not be correct but will merely be congruent to the correct solutions. If this condition arises we say we have *pseudo-overflow.**

---

*This term is due to T. M. Rao. We use the term pseudo-overflow rather than overflow because if it occurs in computing an intermediate result, it creates no difficulty, and if it occurs in computing a final result, the correct answer usually can be obtained with additional information.

2.30. **Remark.** In Definition 2.15 we defined the multiplicative inverse of an integer b modulo m. Subsequently, we gave necessary and sufficient conditions for its existence and uniqueness, and we gave examples illustrating its use. However, no algorithm has been presented here for computing it. The reason for this is that a simple and practical algorithm is presented in Section 5 where its appearance seems more appropriate.

EXERCISES I.2

1. Evaluate
   (a) $|279|_5$
   (b) $|-279|_5$
   (c) $|10^6|_{37}$
   (d) $|48|_{11}$

   (e) $/279/_5$
   (f) $/-279/_5$
   (g) $/10^6/_{37}$
   (h) $/48/_{11}$

2. Demonstrate four ways to evaluate $|27 \cdot 17|_{13}$.

3. Which elements in $\mathbb{I}_{16}$ have multiplicative inverses modulo 16? Find those multiplicative inverses which exist.

3 Multiple-Modulus Residue Arithmetic

The best way to handle the problem of pseudo-overflow, described in Remark 2.29, is to use more than one modulus. This is due to the fact that multiple-modulus residue arithmetic using several moduli can be shown to be equivalent to single-modulus residue arithmetic using the least common multiple of the moduli as the single modulus.

For example, consider the ordered n-tuple

$$(3.1) \qquad \beta = [m_1, m_2, \ldots, m_n],$$

whose components are the (distinct) moduli $m_1, m_2, \ldots, m_n$. Assume that the moduli are pairwise relatively prime, that is, that

$$(3.2) \qquad \gcd(m_i, m_j) = 1 \qquad i \neq j.$$

Whenever β in (3.1) satisfies (3.2) we call it a *base vector* for the residue number systems we are about to describe.

3.3. **Definition.** For each integer s we call the (unique) ordered n-tuple of residues

$$|s|_\beta = [|s|_{m_1}, |s|_{m_2}, \ldots, |s|_{m_n}]$$

the *standard* residue representation* of s with respect to the base vector β. The individual residues $|s|_{m_i}$ are called the *standard residue digits* of s with respect to β.

---

* We use the word "standard" to distinguish these representations from the "symmetric" residue representations mentioned in Remark 3.14.

3.4. EXAMPLE. Let $\beta = [5, 7, 9]$ and $s = 34$. Then

$$|34|_\beta = [4, 6, 7].$$

Let M be the product of the moduli in the base vector β, that is, let

$$(3.5) \qquad\qquad M = \prod_{i=1}^{n} m_i.$$

We are now in a position to state an important theorem and its corollary. In both the theorem and the corollary M should be replaced by the least common multiple of the moduli. However, since the moduli are assumed to be pairwise relatively prime, M is the least common multiple.

3.6. **Theorem.** *Two integers s and t have the same standard residue representation with respect to β, that is, $|t|_\beta = |s|_\beta$, if and only if,*

$$s \equiv t \qquad (\mathrm{mod}\ M).$$

3.7. **Corollary.** *If $t = |s|_M$, then t and s have the same standard residue representation with respect to β, that is, $|t|_\beta = |s|_\beta$.*

3.8. EXAMPLE. Let $\beta = [3, 5, 7]$ so that $M = 105$. Also, let $s = 403$. Then

$$|403|_{105} = 88.$$

It is easily verified (by direct computation) that

$$|403|_\beta = |88|_\beta$$
$$= [1, 3, 4].$$

Suppose we consider, for each $a \in \mathbb{I}$, the unique integer r which satisfies

$$(3.9) \qquad\qquad a \equiv r \qquad (\mathrm{mod}\ M)$$

and

$$(3.10) \qquad\qquad 0 \le r < M.$$

Then

$$(3.11) \qquad\qquad r = |a|_M$$

belongs to the set of non-negative residues modulo M

$$(3.12) \qquad\qquad \mathbb{I}_M = \{0, 1, 2, \ldots, M - 1\}.$$

3.13. **Definition.** *The standard residue number system* for the base vector β is the M-member set of standard residue representations

$$\mathbb{I}_\beta = \{|s|_\beta : s \in \mathbb{I}\}.$$

Obviously, $\mathbb{I}_\beta$ is a finite number system and its elements are in one-to-one correspondence with the elements of $\mathbb{I}_M$ from Theorem 3.6 and Corollary 3.7.

As a consequence of Theorem 2.12, the system $(\mathbb{I}_M, +, \cdot)$, where $+$ and $\cdot$ denote addition and multiplication modulo M, constitutes a finite commutative ring. It is not difficult to show that we can define binary operations $\boxplus$ and $\boxdot$ on elements of $\mathbb{I}_\beta$ in such a way that the system $(\mathbb{I}_\beta, \boxplus, \boxdot)$ also constitutes a finite commutative ring. (See Theorem 3.15 and Theorem 3.16, for example).

It can be shown that the rings $(\mathbb{I}_M, +, \cdot)$ and $(\mathbb{I}_\beta, \boxplus, \boxdot)$ are isormorphic and, as a consequence, multiple-modulus residue arithmetic is equivalent to single-modulus residue arithmetic with M as the modulus.

3.14. Remark. If $\beta = [m_1, m_2, \ldots, m_n]$ contains only odd moduli, we can use (2.23) and Definition 2.24 to define the corresponding *symmetric residue number system* for the base vector β. In this case we represent an integer s by the *symmetric residue representation*.

$$/s/_\beta = [/s/_{m_1}, /s/_{m_2}, \ldots, /s/_{m_n}],$$

where the individual residues $/s/_{m_i}$ are the *symmetric residue digits*. We define the symmetric residue number system to be the M-member set of symmetric residue representations

$$\mathbb{S}_\beta = \{/s/_\beta : s \in \mathbb{I}\}.$$

In $\mathbb{S}_\beta$ every integer s is represented by a unique n-tuple $/s/_\beta$ and the correspondence is one-to-one for those integers in the set

$$\mathbb{S}_M = \left\{ -\frac{M-1}{2}, \ldots, -2, -1, 0, 1, 2, \ldots, \frac{M-1}{2} \right\}.$$

Analogous to $(\mathbb{I}_M, +, \cdot)$ and $(\mathbb{I}_\beta, \boxplus, \boxdot)$ we have the finite commutative rings $(\mathbb{S}_M, +, \cdot)$ and $(\mathbb{S}_\beta, \boxplus, \boxdot)$. It can be shown that they are isomorphic and, indeed, that the latter two rings are isomorphic with the former two. Thus, $\mathbb{I}_\beta$, $\mathbb{S}_\beta$, $\mathbb{I}_M$, and $\mathbb{S}_M$ are all M-member sets and we have a one-to-one correspondence between the standard and the symmetric residue number systems. See Table 3.26 on p. 16.

Note that for any integer a, conversions between $|a|_\beta$ and $/a/_\beta$ are accomplished by the mappings (2.25) and (2.26) applied to corresponding residue digits $|a|_{m_i}$ and $/a/_{m_i}$ for $i = 1, 2, \ldots, n$.

Arithmetic in $\mathbb{I}_\beta$

We describe addition, subtraction, and multiplication in $\mathbb{I}_\beta$ as follows:

3.15. Theorem. *Let a and b be integers. The standard residue representation of the integer $a \pm b$ with respect to the base vector β is the ordered n-tuple*

$$|a \pm b|_\beta = [z_1, z_2, \ldots, z_n]$$

where

$$z_i = \left||a|_{m_i} \pm |b|_{m_i}\right|_{m_i} \qquad i = 1, 2, \ldots, n.$$

3.16. Theorem. *Let a and b be integers. The standard residue representation of the integer ab with respect to the base vector β is the ordered n-tuple*

$$|ab|_\beta = [w_1, w_2, \ldots, w_n]$$

where

$$w_i = \left||a|_{m_i}|b|_{m_i}\right|_{m_i} \qquad i = 1, 2, \ldots, n.$$

Thus, to get the residue digits of a sum, a difference, or a product of two integers we merely add, subtract, or multiply, respectively, the corresponding residue digits of the two operands in *componentwise fashion* and reduce each result modulo the appropriate modulus.

3.17. Example. Let $\beta = [3, 5, 7]$ with $M = 105$. Also, let $a = 24$ and $b = 20$. Then

$$|24|_\beta = [0, 4, 3]$$

and

$$|20|_\beta = [2, 0, 6].$$

Thus,

$$|24 + 20|_\beta = [|0 + 2|_3, |4 + 0|_5, |3 + 6|_7]$$
$$= [2, 4, 2].$$

Similarly,

$$|24 - 20|_\beta = [|0 - 2|_3, |4 - 0|_5, |3 - 6|_7]$$
$$= [1, 4, 4],$$

and

$$|(24)(20)|_\beta = [|(0)(2)|_3, |(4)(0)|_5, |(3)(6)|_7]$$
$$= [0, 0, 4].$$

As a check we observe that

$$|44|_\beta = [2, 4, 2],$$
$$|4|_\beta = [1, 4, 4],$$
$$|480|_\beta = [0, 0, 4].$$

3.18. Remark. In the three computations just described notice that the correct sum, difference, and product of 24 and 20 are 44, 4, and 480, respectively.

The first two results are elements of the set $\mathbb{I}_{105}$ but the third is not. Thus, the representations $[2, 4, 2]$ and $[1, 4, 4]$ will be correctly mapped* onto 44 and 4, respectively, but $[0, 0, 4]$ will be mapped onto 60 (not 480) since

$$0 \le 60 < 105$$

and

$$480 \equiv 60 \qquad (\text{mod } 105).$$

Notice that $M = 105$ was not large enough to prevent pseudo-overflow in the case of the product. The only thing to do, of course, is to make M larger either by choosing *more* moduli or by choosing *larger* moduli.

In order to do division in the standard residue number system we have a slight generalization of the procedure used in single-modulus residue arithmetic.

3.19. **Definition.** Let a be an integer and let $a^{-1}(m_1), a^{-1}(m_2), \ldots, a^{-1}(m_n)$ exist. Then the n-tuple

$$a^{-1}(\beta) = [a^{-1}(m_1), a^{-1}(m_2), \ldots, a^{-1}(m_n)]$$

is called the *standard residue representation of the multiplicative inverse of a with respect to the base vector β.*

For the existence of $a^{-1}(\beta)$ we merely require that $a^{-1}(m_i)$ exist for $i = 1, 2, \ldots, n$. Obviously, when $a^{-1}(\beta)$ exists, it is *unique*.

3.20. **Theorem.** *Let a and b be integers and let $a^{-1}(\beta)$ exist. Then*

$$\left| \frac{b}{a} \right|_\beta = [c_1, c_2, \ldots, c_n],$$

where

$$c_i = \left| |b|_{m_i} a^{-1}(m_i) \right|_{m_i} \qquad i = 1, 2, \ldots, n.$$

Notice that

(3.21)
$$\left| \frac{a}{a} \right|_\beta = [1, 1, \ldots, 1],$$

as it should.

3.22. EXAMPLE. Let $\beta = [3, 5, 7]$ with $M = 105$, and let $a = 23$ and $b = 46$. In this example a divides b. Consequently, since

$$|46|_\beta = [1, 1, 4]$$

and

$$23^{-1}(\beta) = [2, 2, 4],$$

---

*The method used to map residue representations onto integers is described in the next section.

we have

$$|46/23|_\beta = [|1 \cdot 2|_3, |1 \cdot 2|_5, |4 \cdot 4|_7]$$
$$= [2, 2, 2],$$

and this is the correct representation for 2.

3.23. Remark. If a does not divide b we have a result analogous to the situation associated with single-modulus residue arithmetic. The answer may be difficult to interpret (without further information) but it is valid as an intermediate result for further computation. See Examples 2.20 and 2.21.

3.24. Remark. It needs to be emphasized that even though multiple-modulus residue arithmetic is equivalent to single-modulus residue arithmetic (with modulus M), there is one important difference. Since m can always be chosen to be a prime, we can guarantee that $(\mathbb{I}_m, +, \cdot)$ is a finite field. On the other hand, M cannot be a prime (by definition) and so $(\mathbb{I}_M, +, \cdot)$ can only be a finite commutative ring. Consequently, it is to be expected that there may be problems with division in multiple-modulus residue arithmetic because $a^{-1}(\beta)$ will not exist for every base vector β and every integer a. See Theorems 7.16 and 7.17 in Section 7.

3.25. Remark. For a description of multiple-modulus residue arithmetic applied to the solution of systems of linear algebraic equations, see Howell and Gregory [1969], [1970] and Young and Gregory [1973], Chapter 13.

3.26. Table. Representations for Integers, Where $\beta = [3, 5]$ and $M = 15$.

$\mathbb{I}_M$	$\mathbb{S}_M$	$\mathbb{I}_\beta$	$\mathbb{S}_\beta$
0	0	[0, 0]	[0, 0]
1	1	[1, 1]	[1, 1]
2	2	[2, 2]	[−1, 2]
3	3	[0, 3]	[0, −2]
4	4	[1, 4]	[1, −1]
5	5	[2, 0]	[−1, 0]
6	6	[0, 1]	[0, 1]
7	7	[1, 2]	[1, 2]
8	−7	[2, 3]	[−1, −2]
9	−6	[0, 4]	[0, −1]
10	−5	[1, 0]	[1, 0]
11	−4	[2, 1]	[−1, 1]
12	−3	[0, 2]	[0, 2]
13	−2	[1, 3]	[1, −2]
14	−1	[2, 4]	[−1, −1]

EXERCISES I.3

1. Let $\beta = [3, 5, 7]$. Give the representations for
   (a) $|137|_\beta$ (d) $/137/_\beta$
   (b) $|-137|_\beta$ (e) $/-137/_\beta$
   (c) $|537|_\beta$ (f) $/537/_\beta$.

2. Let $\beta = [5, 7, 11]$, $a = 34$, and $b = 408$. Find the representation for
   (a) $|a + b|_\beta$ (e) $/a + b/_\beta$
   (b) $|a - b|_\beta$ (f) $/a - b/_\beta$
   (c) $|ab|_\beta$ (g) $/ab/_\beta$
   (d) $\left|\dfrac{b}{a}\right|_\beta$ (h) $\left/\dfrac{b}{a}\right/_\beta$.

3. Which results involve "pseudo-overflow" in the previous problem?

4. Prove Theorem 3.6.

5. Prove Corollary 3.7.

6. Prove Theorem 3.15.

7. Prove Theorem 3.16.

8. Prove Theorem 3.20.

4 Mapping Standard Residue Representations onto Integers

In Remark 3.18 we indicated that, for $\beta = [3, 5, 7]$, the residue representations $[2, 4, 2]$, $[1, 4, 4]$, and $[0, 0, 4]$ were mapped onto the integers 44, 4, and 60, respectively, but we gave no indication as to how the mappings were carried out. The purpose of this section is to answer the question "how do we map a standard residue representation (with respect to β) onto a *unique* integer in $\mathbb{I}_M$?"

One of the oldest known algorithms (but not the fastest) for carrying out this mapping makes use of a classical theorem from the theory of numbers called the *Chinese Remainder Theorem*. (See Young and Gregory [1973], p. 874, for example.) We shall not describe that algorithm here, however. Instead, we present an algorithm which uses a *mixed-radix number representation* for each integer.

Consider the ordered n-tuple of integers

(4.1) $$\rho = [r_1, r_2, \ldots, r_n]$$

where the components $r_1, r_2, \ldots, r_n$ are called *radices*. Let R be the product of the radices, that is,

(4.2) $$R = \prod_{i=1}^{n} r_i.$$

It is well known (see Szabó and Tanaka [1967], p. 41, for example) that every integer s in the range

(4.3) $$0 \leq s < R$$

can be expressed uniquely in the form

(4.4) $$s = d_0 + d_1(r_1) + d_2(r_1 r_2) + \cdots + d_{n-1}(r_1 r_2 \cdots r_{n-1}),$$

where $d_0, d_1, \ldots, d_{n-1}$ are the *standard mixed-radix digits* satisfying the inequalities

(4.5) $$0 \leq d_i < r_{i+1} \qquad i = 0, 1, \ldots, n - 1.$$

Notice that the primary role played by r_n is to establish a bound on d_{n-1}.

The *digit sequence* for s in this mixed-radix representation is the ordered set of digits $d_0, d_1, \ldots, d_{n-1}$ which we display in the form

(4.6) $$\langle s \rangle_\rho = \langle d_0, d_1, \ldots, d_{n-1} \rangle.$$

For example, if $\rho = [2, 3, 5]$, then $R = 30$. Consequently, since

(4.7) $$29 = 1 + 2(2) + 4(2 \cdot 3),$$

we know that $d_0 = 1$, $d_1 = 2$, and $d_2 = 4$. Hence,

(4.8) $$\langle 29 \rangle_\rho = \langle 1, 2, 4 \rangle.$$

4.9. Definition. The *standard mixed-radix system* for ρ is the set of digit sequences $\langle s \rangle_\rho$ for integers in the range

$$0 \leq s < R.$$

A special case occurs if $r_1 = r_2 = \cdots = r_n$, the familiar *fixed-radix number representation*. If each radix is ten, for example, this is merely the decimal representation. A more interesting special case, (for our purposes) occurs if $r_i = m_i$ for $i = 1, 2, \ldots, n$, where m_i is an element of the base vector β for our multiple-modulus residue system, and r_i is an element of ρ defined in (4.1). In this case

(4.10) $$\rho = \beta,$$

and so the ranges for the multiple-modulus residue system and its associated mixed-radix system are both $0 \leq s < M$. This is extremely important since we shall wish to change from one system to the other.

Consider the standard mixed-radix system and its associated standard residue system for the base vector $\beta = [m_1, m_2, \ldots, m_n]$. If an integer s has the representation

(4.11) $$s = d_0 + d_1(m_1) + d_2(m_1 m_2) + \cdots + d_{n-1}(m_1 m_2 \ldots m_{n-1}),$$

with the (unique) digit sequence

(4.12) $$\langle s \rangle_\beta = \langle d_0, d_1, \ldots, d_{n-1} \rangle$$

in the former, and the residue representation

(4.13)
$$|s|_\beta = [|s|_{m_1}, |s|_{m_2}, \ldots, |s|_{m_n}]$$

in the latter, we see from (4.5) and (2.6) that both the mixed-radix digits d_{i-1} and the residue digits $|s|_{m_i}$ lie in the same closed interval $[0, m_i - 1]$ for $i = 1, 2, \ldots, n$.

Suppose we are given the residue representation $|s|_\beta$ in (4.13) and we wish to find $\langle s \rangle_\beta$ in (4.12). In other words, suppose we are given the residue digits $|s|_{m_i}$ and we wish to find the mixed-radix digits d_{i-1} for $i = 1, 2, \ldots, n$. To get d_0 let $s = t_1$ and observe that, from (4.11).

$$t_1 = s$$

(4.14)
$$= d_0 + m_1[d_1 + d_2(m_2) + \cdots + d_{n-1}(m_2 \ldots m_{n-1})]$$
$$= d_0 + m_1 t_2.$$

Hence, from Theorem 2.10,

(4.15)
$$|t_1|_{m_1} = |d_0 + m_1 t_2|_{m_1}$$
$$= d_0.$$

Notice that $d_0 = |t_1|_{m_1} = |s|_{m_1}$, which implies that the first mixed-radix digit equals the first residue digit and *no computation is necessary*.

To compute d_1 we use (4.14) and write

(4.16)
$$t_2 = d_1 + m_2[d_2 + d_3(m_3) + \cdots + d_{n-1}(m_3 \ldots m_{n-1})]$$
$$= d_1 + m_2 t_3.$$

Hence, from Theorem 2.10,

(4.17)
$$|t_2|_{m_2} = |d_1 + m_2 t_3|_{m_2}$$
$$= d_1.$$

This suggests the following recursion for obtaining the mixed-radix digits. Beginning with the initial values $t_1 = s$ and $d_0 = |t_1|_{m_1}$ we successively compute t_{i+1} and d_i (in that order) using the equations

(4.18)
$$\begin{cases} t_{i+1} = \dfrac{t_i - d_{i-1}}{m_i} \\ d_i = |t_{i+1}|_{m_{i+1}} \end{cases} \quad i = 1, 2, \ldots, n - 1.$$

The computation indicated in (4.18) for computing the mixed-radix digits can be carried out using residue arithmetic. This is a critical point because *it cannot be carried out using ordinary arithmetic*. The reason for this is that we must know s to get started (recall our definition, $t_1 = s$) and we do not know s. (That is what we are trying to find.) However, we do know $|s|_\beta$ and that is all we need when we use residue arithmetic.

The computation of d_i using residue arithmetic

Let $|s|_\beta$ be defined by (4.13). If we set $s = t_1$ and use (4.15) we can write

$$(4.19) \qquad |t_1|_\beta = [d_0, |t_1|_{m_2}, \ldots, |t_1|_{m_n}].$$

By definition,

$$(4.20) \qquad |d_0|_\beta = [|d_0|_{m_1}, |d_0|_{m_2}, \ldots, |d_0|_{m_n}],$$

which allows us to compute

$$(4.21) \qquad |t_1 - d_0|_\beta = [0, |z_2|_{m_2}, |z_3|_{m_3}, \ldots, |z_n|_{m_n}],$$

with

$$(4.22) \qquad z_i = |t_1|_{m_i} - |d_0|_{m_i} \qquad i = 2, 3, \ldots, n.$$

If we introduce the reduced base vector

$$(4.23) \qquad \beta_1 = [m_2, m_3, \ldots, m_n],$$

we can represent $t_1 - d_0$ with respect to β_1. In this case,

$$(4.24) \qquad |t_1 - d_0|_{\beta_1} = [|z_2|_{m_2}, |z_3|_{m_3}, \ldots, |z_n|_{m_n}].$$

In order to compute t_2 we need $m_1^{-1}(\beta_1)$. This multiplicative inverse exists because m_1 is relatively prime to each of the elements of β_1. Thus,

$$(4.25) \qquad m_1^{-1}(\beta_1) = [m_1^{-1}(m_2), m_1^{-1}(m_3), \ldots, m_1^{-1}(m_n)]$$

and so

$$(4.26) \qquad \begin{aligned} |t_2|_{\beta_1} &= |(t_1 - d_0)/m_1|_{\beta_1} \\ &= [|w_2|_{m_2}, |w_3|_{m_3}, \ldots, |w_n|_{m_n}], \end{aligned}$$

with

$$(4.27) \qquad w_i = |z_i|_{m_i} m_1^{-1}(m_i) \qquad i = 2, 3, \ldots, n.$$

From (4.17) and (4.26) we have

$$(4.28) \qquad \begin{aligned} |w_2|_{m_2} &= |t_2|_{m_2} \\ &= d_1, \end{aligned}$$

the second mixed-radix digit.

If we use this result in (4.26) and note that $|w_i|_{m_i} = |t_2|_{m_i}$, we can write

$$(4.29) \qquad |t_2|_{\beta_1} = [d_1, |t_2|_{m_3}, \ldots, |t_2|_{m_n}].$$

Analogous to (4.20) we have

$$(4.30) \qquad |d_1|_{\beta_1} = [|d_1|_{m_2}, |d_1|_{m_3}, \ldots, |d_1|_{m_n}]$$

which gives us

$$(4.31) \qquad |t_2 - d_1|_{\beta_1} = [0, |v_3|_{m_3}, \ldots, |v_n|_{m_n}],$$

with

(4.32) $$v_i = |t_2|_{m_i} - |d_1|_{m_i} \qquad i = 3, 4, \ldots, n.$$

If, analogous to β_1, we introduce the reduced base vector

(4.33) $$\beta_2 = [m_3, m_4, \ldots, m_n],$$

we can represent $t_2 - d_1$ with respect to β_2. In this case,

(4.34) $$|t_2 - d_1|_{\beta_2} = [|v_3|_{m_3}, |v_4|_{m_4}, \ldots, |v_n|_{m_n}].$$

In order to compute t_3 we need $m_2^{-1}(\beta_2)$. This multiplicative inverse exists because m_2 is relatively prime to each of the elements of β_2. Thus,

(4.35) $$m_2^{-1}(\beta_2) = [m_2^{-1}(m_3), m_2^{-1}(m_4), \ldots, m_2^{-1}(m_n)]$$

and so

(4.36) $$\begin{aligned} |t_3|_{\beta_2} &= |(t_2 - d_1)/m_2|_{\beta_2} \\ &= [|u_3|_{m_3}, |u_4|_{m_4}, \ldots, |u_n|_{m_n}], \end{aligned}$$

with

(4.37) $$u_i = |v_i|_{m_i} m_2^{-1}(m_i) \qquad i = 3, 4, \ldots, n.$$

From (4.18) we know that

(4.38) $$\begin{aligned} |u_3|_{m_3} &= |t_3|_{m_3} \\ &= d_2, \end{aligned}$$

the third mixed-radix digit.

If we continue this algorithm, we eventually compute each of the mixed-radix digits in sequence.

4.39. PROBLEM. If $\beta = [13, 11, 7]$ and $|s|_\beta = [4, 2, 4]$, what is $\langle s \rangle_\beta$ and what is s?

SOLUTION. We assume, since $M_\beta = 1001$, that we seek the unique integer in the range $0 \le s < 1001$. The computation described in (4.19) through (4.38) can be presented in the form of a table.

β	$m_1 = 13$	$m_2 = 11$	$m_3 = 7$	
$\lvert t_1 \rvert_\beta$	4	2	4	
$\lvert d_0 \rvert_\beta$	4	4	4	subtract
$\lvert t_1 - d_0 \rvert_\beta$	0	9	0	
$m_1^{-1}(\beta_1)$		6	6	multiply
$\lvert t_2 \rvert_{\beta_1} = \lvert(t_1 - d_0)/m_1\rvert_{\beta_1}$		10	0	
$\lvert d_1 \rvert_{\beta_1}$		10	3	subtract
$\lvert t_2 - d_1 \rvert_{\beta_1}$		0	4	
$m_2^{-1}(\beta_2)$			2	multiply
$\lvert t_3 \rvert_{\beta_2} = \lvert(t_2 - d_1)/m_2\rvert_{\beta_2}$			1	

The elements in the dotted squares are d_0, d_1, and d_2, in that order. Hence, $\langle s \rangle_\beta = \langle 4, 10, 1 \rangle$. Consequently, from (4.11), we obtain

$$s = 4 + 10(13) + 1(13)(11)$$
$$= 277,$$

and this is the correct answer. □

4.40. Remark. Since we know how to find the digit sequence $\langle s \rangle_\beta$ from $|s|_\beta$ we merely evaluate the right hand side of (4.11) to find the integer s. In the example above this was quite simple to do directly. However, we use the following recursive procedure to evaluate (4.11), in general.

$$s_1 = d_{n-1}$$
$$s_2 = d_{n-2} + s_1 m_{n-1}$$
$$s_3 = d_{n-3} + s_2 m_{n-2}$$
$$\vdots$$
$$s_n = d_0 + s_{n-1} m_1$$

and $s = s_n$.

4.41. Remark. At the beginning of Section 3 it was pointed out that multiple-modulus residue arithmetic using the base vector β is equivalent to single-modulus residue arithmetic using M as the single modulus.* In other words, arithmetic in $(\mathbb{I}_\beta, \boxplus, \boxdot)$ is equivalent to arithmetic in $(\mathbb{I}_M, +, \cdot)$.

Suppose we want to do arithmetic on both positive and negative integers. We use the same procedure suggested in Section 2 where integers in $\mathbb{S}_m$ were mapped into $\mathbb{I}_m$ using (2.25) so that arithmetic could be performed in $(\mathbb{I}_m, +, \cdot)$. Results were then mapped back into $\mathbb{S}_m$ using (2.26).

We can do the same thing here. Integers in $\mathbb{S}_M$ can be mapped into $\mathbb{I}_M$ using a mapping similar to (2.25). This allows the arithmetic to be done in $(\mathbb{I}_\beta, \boxplus, \boxdot)$ producing results in $\mathbb{I}_M$. These results can then be mapped into $\mathbb{S}_M$ using a mapping similar to (2.26).

4.42. Remark. We have discussed the four basic arithmetic operations in multiple-modulus residue arithmetic and conversions between residue representations and integers. Notice that arithmetic is done in a componentwise fashion. This suggests that we could take advantage of a computer especially designed for multiple-modulus residue arithmetic. If we had n processors, each designed to carry out residue operations on integers, the operations on the components could be done simultaneously. The newest generation of computers with their vector operations and parallel processors are a

---

*Actually it is the least common multiple of the moduli which is the single modulus. However, since we are assuming that the moduli are pairwise relatively prime, M is the least common multiple.

step in the right direction. However, none of them has the hardware capability for reducing a result modulo m. It must be done with software.

4.43. Remark. We have shown that addition, subtraction and multiplication are extremely simple operations in multiple-modulus residue arithmetic, but that division is a bit more complicated (see Remark 3.24). There are three other situations which present us with a bit of complexity:

(i) magitude comparison for two integers s and t,
(ii) sign detection of s, and
(iii) the recovery of s from $|s|_\beta$.

For a discussion of (i) and (ii) the reader is referred to Szabó and Tanaka [1967], Chapter 4. We have dealt with (iii) in this section and we have discussed (ii) in Remark 4.41.

EXERCISES I.4

1. If $\beta = [3, 5, 2]$,
    (a) What is $\langle 29 \rangle_\beta$?
    (b) If $\langle s \rangle_\beta = \langle 1, 2, 3 \rangle$, what is s?

2. If $\beta = [5, 7, 3]$ and $|s|_\beta = [0, 3, 1]$, what is $\langle s \rangle_\beta$ and what is s?

3. If $\beta = [3, 7, 5]$ and $|s|_\beta = [1, 2, 0]$, what is $\langle s \rangle_\beta$ and what is s?

5 Single-Modulus Residue Arithmetic with Rational Numbers

It turns out that we can use single-modulus residue arithmetic to (in effect) carry out arithmetic operations on certain rational numbers. The basic idea is to map the rational operands into the set of integers $\mathbb{I}_m$ defined in (2.6), carry out the arithmetic operations in $(\mathbb{I}_m, +, \cdot)$ and then map the integer results back onto the appropriate rational numbers.

For this application we shall find it useful to choose our modulus to have the form $m = p^r$, where p is a prime and r is a positive integer. We then define a mapping of elements a/b, for which $\gcd(b, p) = 1$, into $\mathbb{I}_m$ by making use of the fact that $b^{-1}(m)$ exists if and only if $\gcd(b, p) = 1$. Obviously, if $r = 1$, m is the prime p.

5.1. Definition. If $x = a/b$, and if $\gcd(b, p) = 1$ so that $b^{-1}(m)$ exists, then

$$|x|_m = \left| \frac{a}{b} \right|_m$$
$$= |ab^{-1}|_m.$$

When we recall Definition 2.19, it is clear that we are now interpreting Example 2.20 as a mapping of the rational number $\frac{7}{9}$ onto the integer 2 in $\mathbb{I}_{11}$.

Let $\hat{\mathbb{Q}}$ be the set of rational numbers which can be mapped onto $\mathbb{I}_m$ by this mapping, that is, let

$$(5.2) \qquad \hat{\mathbb{Q}} = \left\{ \frac{a}{b} : \gcd(b, p) = 1 \right\}.$$

Then, in Definition 5.1 we are describing the mapping $|\cdot|_m : \hat{\mathbb{Q}} \to \mathbb{I}_m$.

It turns out that each integer $k \in \mathbb{I}_m$ is the image of an infinite set of elements of $\hat{\mathbb{Q}}$ which we shall label $\mathbb{Q}_k$. Thus, for $k = 0, 1, \ldots, m - 1$,

$$(5.3) \qquad \mathbb{Q}_k = \left\{ \frac{a}{b} \in \hat{\mathbb{Q}} : \left|\frac{a}{b}\right|_m = k \right\},$$

from which it follows that

$$(5.4) \qquad \hat{\mathbb{Q}} = \bigcup_{k=0}^{m-1} \mathbb{Q}_k.$$

The set $\mathbb{Q}_0$ (the rational numbers in $\mathbb{Q}$ which are mapped onto zero) consists of those numbers a/b, with $\gcd(b, p) = 1$, for which $|a|_m = 0$. We call the disjoint subsets $\mathbb{Q}_0, \mathbb{Q}_1, \ldots, \mathbb{Q}_{m-1}$, *generalized residue classes* modulo m, because they contain the ordinary residue classes of integers, defined in (2.8), as proper subsets, that is,

$$(5.5) \qquad \mathbb{R}_k \subset \mathbb{Q}_k, \qquad k = 0, 1, 2, \ldots, m - 1.$$

The following theorem and corollary characterize the elements of a generalized residue class.

5.6. **Theorem.** *Let $x = a/b$ and $y = c/d$, where both $b^{-1}(m)$ and $d^{-1}(m)$ exist. Then*

$$|x|_m = |y|_m$$

if and only if

$$ad \equiv bc \qquad (\mathrm{mod}\ m).$$

PROOF. Assume that $ad \equiv bc$ (mod m). If we multiply both sides of the congruence by $b^{-1}d^{-1}$, we obtain

$$ab^{-1} \equiv cd^{-1} \qquad (\mathrm{mod}\ m)$$

which implies

$$|ab^{-1}|_m = |cd^{-1}|_m.$$

Thus, from Definition 5.1, we have

$$|x|_m = |y|_m.$$

We leave the proof of the converse to the reader.   □

5.7. Corollary. *Let* $x = a/b$ *and* $y = c/d$ *be elements of* $\hat{\mathbb{Q}}$. *Then* x *and* y *belong to the same generalized residue class* $\mathbb{Q}_k$ *if and only if*

$$ad \equiv bc \qquad (\text{mod } m).$$

PROOF. By definition, two rational numbers x and y belong to the same generalized residue class if and only if

$$|x|_m = |y|_m,$$

and this equality holds if and only if

$$ad \equiv bc \qquad (\text{mod } m)$$

by Theorem 5.6. The result follows. $\qquad\qquad\qquad\qquad\qquad\qquad\square$

Not every rational number a/b satisfies the condition $\gcd(b, p) = 1$, and so not every rational number is an element of $\hat{\mathbb{Q}}$, that is, of one of the disjoint subsets $\mathbb{Q}_0, \mathbb{Q}_1, \ldots, \mathbb{Q}_{m-1}$ which are mapped onto $\mathbb{I}_m$.

5.8. Remark. In view of the fact that $ka/kb = a/b$ for every integer $k \neq 0$, we assume throughout this discussion that *all rational numbers are in reduced form*. Thus, if $x = 2p/3p$, and $p > 3$, we know that $|x|_m$ exists because $x = \frac{2}{3}$. Under this assumption, then, the only rational number $x = a/b$ for which $|x|_m$ does not exist is a rational number that has a denominator which is an integral multiple of p.

5.9. Lemma. *Let* $x = a/b$ *and* $y = c/d$ *be elements of* $\hat{\mathbb{Q}}$. *Then* $x + y$ *and* xy *are elements of* $\hat{\mathbb{Q}}$.

PROOF. If $b \neq 0$ and $d \neq 0$ are not integral multiples of p, then $bd \neq 0$ is not an integral multiple of p. Hence both $xy = ac/bd$ and $x + y = (ad + bc)/bd$ are elements of $\hat{\mathbb{Q}}$. $\qquad\qquad\qquad\qquad\qquad\square$

5.10. Theorem. *The system* $(\hat{\mathbb{Q}}, +, \cdot)$ *is a commutative ring with identity.*

PROOF. From Lemma 5.9, $\hat{\mathbb{Q}}$ is closed under addition and multiplication. Also, $\hat{\mathbb{Q}}$ inherits the commutative and associative laws for both addition and multiplication, and the distributive laws from $\mathbb{Q}$. Finally, it is easy to verify that both 0 and 1 are elements of $\hat{\mathbb{Q}}$, and that the additive inverses of all elements of $\hat{\mathbb{Q}}$ are in $\hat{\mathbb{Q}}$. $\qquad\qquad\qquad\qquad\qquad\square$

We now prove the following lemma which establishes a relationship between the ring $(\hat{\mathbb{Q}}, +, \cdot)$ and the ring $(\mathbb{I}_m, +, \cdot)$.

5.11. Lemma. *Let* $x = a/b$ *and* $y = c/d$ *both belong to* $\hat{\mathbb{Q}}$. *Then*

(i)
$$\left| |x|_m \cdot |y|_m \right|_m = |xy|_m$$

and

(ii)
$$\left| \, |x|_m + |y|_m \, \right|_m = |x + y|_m .$$

PROOF.
$$\left| \, |x|_m \cdot |y|_m \, \right|_m = \left| \, |ab^{-1}|_m \cdot |cd^{-1}|_m \, \right|_m$$
$$= |ab^{-1}cd^{-1}|_m$$
$$= |ac(bd)^{-1}|_m$$
$$= |xy|_m .$$

Likewise,

$$\left| \, |x|_m + |y|_m \, \right|_m = \left| \, |ab^{-1}|_m + |cd^{-1}|_m \, \right|_m$$
$$= |ab^{-1} + cd^{-1}|_m$$
$$= |ab^{-1}dd^{-1} + cd^{-1}bb^{-1}|_m$$
$$= |(ad + bc)(bd)^{-1}|_m$$
$$= |x + y|_m . \qquad \square$$

This lemma establishes the following fundamental result.

5.12. Theorem. *The mapping $|\cdot|_m : \hat{\mathbb{Q}} \to \mathbb{I}_m$ is a homomorphism with respect to addition and multiplication.*

In other words, $\mathbb{I}_m$ is a homomorphic image of $\hat{\mathbb{Q}}$ and so arithmetic operations performed in the ring $(\hat{\mathbb{Q}}, +, \cdot)$ correspond to those same arithmetic operations performed in the ring $(\mathbb{I}_m, +, \cdot)$. Keep in mind that if $r = 1$, $m = p^r$ is the prime p and, in this case, $(\mathbb{I}_p, +, \cdot)$ is a finite field isomorphic to the Galois field $\mathrm{GF}(p)$.

It would be ideal for our purposes if the relationship between $(\hat{\mathbb{Q}}, +, \cdot)$ and $(\mathbb{I}_m, +, \cdot)$ were an isomorphism. However, this calls for a mapping between $\hat{\mathbb{Q}}$ and $\mathbb{I}_m$ which is both one-to-one and onto, and no such mapping exists.

Choosing a useful subset of $\hat{\mathbb{Q}}$

The mapping $|\cdot|_m : \hat{\mathbb{Q}} \to \mathbb{I}_m$ is onto but it is not one-to-one because each integer $k \in \mathbb{I}_m$ is the image of the infinite subset $\mathbb{Q}_k$ of rational numbers. Hence, the mapping does not have an inverse. The question now is whether or not we can identify a unique element in each generalized residue class $\mathbb{Q}_k$ with which we can establish a one-to-one mapping onto the integers in $\mathbb{I}_m$. If this could be done, we would have a one-to-one and onto mapping between these unique elements (in the generalized residue classes $\mathbb{Q}_0$, $\mathbb{Q}_1$, $\ldots$, $\mathbb{Q}_{m-1}$) and their images in $\mathbb{I}_m$, and this mapping would have an inverse.

Unfortunately, we can identify a unique element in only some of the generalized residue classes $\mathbb{Q}_k$, but not all of them. As a consequence, we must settle for a one-to-one and onto mapping between those unique elements which can be found and their images in $\mathbb{I}_m$.

5.13. Definition. The finite subset of $\hat{\mathbb{Q}}$

$$\mathbb{F}_N = \{a/b \in \hat{\mathbb{Q}} : \gcd(a, b) = 1 \text{ and } 0 \le |a| \le N, 0 < |b| \le N\},$$

where $N > 0$ is an integer, is called the set of *order-N Farey fractions*.

It turns out that if we choose N properly, then each generalized residue class $\mathbb{Q}_k$ contains at most one element of $\mathbb{F}_N$.

5.14. Theorem. *Let N be the largest integer satisfying the inequality*

$$2N^2 + 1 \le m$$

and let the generalized residue class $\mathbb{Q}_k$ contain the order-N Farey fraction $x = a/b$. Then x is the only order-N Farey fraction in $\mathbb{Q}_k$.

PROOF. Suppose $x = a/b$ and $y = c/d$ are two order-N Farey fractions contained in $\mathbb{Q}_k$. Then

$$|x|_m = |y|_m = k.$$

From Theorem 5.6, this implies

$$ad \equiv bc \qquad (\mathrm{mod}\ m)$$

or

$$|ad - bc|_m = 0.$$

Furthermore, since

$$0 \le |ad - bc|$$
$$\le |a| \cdot |d| + |b| \cdot |c|$$
$$\le 2N^2$$
$$\le m - 1,$$

it follows that $ad - bc = 0$. But this implies

$$a/b = c/d,$$

or $x = y$. Hence, x is unique. $\qquad\square$

5.15. Remark. If N is chosen as in Theorem 5.14, then the number of elements in $\mathbb{F}_N$ is less than m. Consequently, not every generalized residue class $\mathbb{Q}_0, \mathbb{Q}_1, \ldots, \mathbb{Q}_{m-1}$ can contain an element of $\mathbb{F}_N$. However, if $\mathbb{Q}_k$ does contain an order-N Farey fraction, it contains only one.

We are now in a position to establish a one-to-one and onto mapping between certain elements of $\hat{\mathbb{Q}}$ and certain integers in $\mathbb{I}_m$. Suppose we use

$$(5.16) \qquad \hat{\mathbb{I}}_m = \{|a/b|_m : a/b \in \mathbb{F}_N\}$$

to denote the set of images of the order-N Farey fractions. (Unless otherwise stated we shall assume that N is defined as in Theorem 5.14.) This leads us to the following result.

5.17. **Theorem.** *The mapping* $|\cdot|_m : \mathbb{F}_N \to \hat{\mathbb{I}}_m$ *is one-to-one and onto and thus, has an inverse.*

PROOF. We leave the proof as an exercise for the reader. $\qquad\qquad\qquad\square$

5.18. EXAMPLE. Let $m = 19$. Then $N = 3$ and the mapping $|\cdot|_{19} : \mathbb{F}_3 \to \hat{\mathbb{I}}_{19}$ is exhibited in the following table:

0	0		
1	1	-1	18
2	2	-2	17
3	3	-3	16
.	.	.	.
$-\frac{1}{3}$	6	$\frac{1}{3}$	13
$\frac{2}{3}$	7	$-\frac{2}{3}$	12
$-\frac{3}{2}$	8	$\frac{3}{2}$	11
$-\frac{1}{2}$	9	$\frac{1}{2}$	10

Notice that $\hat{\mathbb{I}}_{19} \subset \mathbb{I}_{19}$ because the integers 4, 5, 14, and 15 are not the images of any elements in $\mathbb{F}_3$.

5.19. **Remark.** We finally have succeeded in establishing a one-to-one and onto mapping between a finite subset of the rational numbers in $\hat{\mathbb{Q}}$ (the so-called order-N Farey fractions) and a (finite) subset of the integers in $\mathbb{I}_m$. Thus, since $\mathbb{F}_N \subset \hat{\mathbb{Q}}$ and $\hat{\mathbb{I}}_m \subset \mathbb{I}_m$, our earlier statement about arithmetic operations in the ring $(\hat{\mathbb{Q}}, +, \cdot)$ being equivalent to arithmetic operations in the ring $(\mathbb{I}_m, +, \cdot)$ can be called into play. If a very large integer m is chosen, the set of order-N Farey fractions is quite large. If $\mathbb{F}_N$ is large enough to contain all of the data describing a problem and all of the answers to the problem, then we can

  (i) map the operands from $\mathbb{F}_N$ into $\hat{\mathbb{I}}_m$,
 (ii) carry out the arithmetic operations in the ring $(\mathbb{I}_m, +, \cdot)$ free of rounding errors, and
(iii) map the integer results back into $\mathbb{F}_N$.

If some of the answers to the problem are not order-N Farey fractions, then we get a situation analogous to the situation mentioned in Remark 2.29 which we call pseudo-overflow.

The following problems demonstrate how this system works. We illustrate both normal computation and computation resulting in pseudo-overflow.

5.20. PROBLEM. Find x if

$$x = \tfrac{1}{3} - \tfrac{2}{3}$$
$$= \tfrac{1}{3} + (-\tfrac{2}{3}).$$

SOLUTION. If we choose $m = 19$, then $N = 3$, and we can use the mapping exhibited in the table in Example 5.18. Thus,

$$|x|_{19} = |\tfrac{1}{3} + (-\tfrac{2}{3})|_{19}$$
$$= |13 + 12|_{19}$$
$$= 6.$$

Since $6 \in \hat{I}_{19}$ we use the inverse mapping to obtain

$$x = -\tfrac{1}{3}$$

and this is the correct answer.   □

5.21. PROBLEM. Find x if

$$x = \tfrac{1}{2} - \tfrac{2}{3}$$
$$= \tfrac{1}{2} + (-\tfrac{2}{3}).$$

SOLUTION. As before, we choose $m = 19$ with $N = 3$. However, in this problem the solution is not an order-3 Farey fraction and we get pseudo-overflow. To see that this occurs, we write

$$|x|_{19} = |\tfrac{1}{2} + (-\tfrac{2}{3})|_{19}$$
$$= |10 + 12|_{19}$$
$$= 3.$$

Since $3 \in \hat{I}_{19}$ we use the inverse mapping to obtain

$$x = 3,$$

and this is incorrect. (The correct solution is $x = -\tfrac{1}{6}$).   □

However, it is interesting to observe that in this last problem

$$|-\tfrac{1}{6}|_{19} = |(-1)6^{-1}|_{19}$$

(5.22)
$$= |(-1)(16)|_{19}$$
$$= 3.$$

which indicates that both the computed answer ($x = 3$) and the correct answer ($x = -\tfrac{1}{6}$) are elements of the same generalized residue class $\mathbb{Q}_3$.

Pseudo-overflow can occur also when the result computed in $(\mathbb{I}_m, +, \cdot)$ is not an element of $\hat{\mathbb{I}}_m$. Thus, in $\mathbb{I}_{19}$, a result which is either 14, 15, 4, or 5 cannot be mapped onto an element of $\mathbb{F}_3$.

5.23. Remark. It turns out that we can use rational operands which are not elements of F_N, and we can produce *intermediate results* which are not elements of F_N, as long as the *final results* are elements of $\mathbb{F}_N$. This is demonstrated in the following example.

$$x = \tfrac{1}{2} - \tfrac{2}{3} - \tfrac{1}{6}$$
$$= \tfrac{1}{2} + (-\tfrac{2}{3}) + (-\tfrac{1}{6}).$$

In $(\mathbb{I}_{19}, +, \cdot)$, we obtain

$$|x|_{19} = |\tfrac{1}{2} + (-\tfrac{2}{3}) + (-\tfrac{1}{6})|_{19}$$
$$= |10 + 12 + 3|_{19}$$
$$= 6.$$

If we use the inverse mapping exhibited in Example 5.18 we obtain

$$x = -\tfrac{1}{3},$$

and this is the correct result. Notice that $-\tfrac{1}{6}$ is not in $\mathbb{F}_3$ and the sum of the first two terms gave us pseudo-overflow in Problem 5.21. However, as an intermediate result, the "incorrect" sum enables us to obtain the correct final result. Recall Example 2.21.

5.24. Remark. It is obvious that we cannot use a table such as the one in Example 5.18 to implement the *forward mapping* $\mathbb{F}_N \rightarrow \hat{\mathbb{I}}_m$ and the *inverse mapping* $\hat{\mathbb{I}}_m \rightarrow \mathbb{F}_N$ in practice. Consequently, in the next section we describe algorithms for implementing both of these mappings. In the process, we encounter the algorithm needed for forming $b^{-1}(m)$ as a special case of the algorithm for the forward mapping.

EXERCISES I.5

1. Complete the proof of Theorem 5.6.

2. Prove Theorem 5.17.

3. Construct a table, similar to the table in Example 5.18, which exhibits the mapping between $\mathbb{F}_5$ and $\hat{\mathbb{I}}_{53}$.

4. Use the finite field $(\mathbb{I}_{53}, +, \cdot)$ to find x, if
   (a) $x = \tfrac{1}{3} - \tfrac{2}{3}$.
   (b) $x = \tfrac{1}{2} + \tfrac{3}{4}$.
   (c) $x = (\tfrac{2}{3})(\tfrac{5}{2})$.
   (d) $x = \tfrac{1}{2} - \tfrac{2}{3} - \tfrac{1}{6}$ (see Remark 5.23).

6 The Forward Mapping and the Inverse Mapping

In this section we describe an algorithm for implementing the forward mapping $\mathbb{F}_N \to \hat{\mathbb{I}}_m$. Thus, given $a/b \in \mathbb{F}_N$ we form $|a/b|_m \in \hat{\mathbb{I}}_m$. This algorithm is based on the Euclidean Algorithm and yields $b^{-1}(m)$ in the special case $a = 1$.

We also describe an algorithm for implementing the inverse mapping $\hat{\mathbb{I}}_m \to \mathbb{F}_N$. It, too, is based on the Euclidean Algorithm and was discovered independently by Kornerup and Krishnamurthy. When the algorithm is applied to an integer $k \in \hat{\mathbb{I}}_m$ several rational numbers in $\mathbb{Q}_k$ are generated, including the unique order-N Farey fraction belonging to $\mathbb{Q}_k$. It is easy to retrieve this Farey fraction since it is the only rational number generated which satisfies Definition 5.13.

The Euclidean Algorithm

First, we consider the Euclidean Algorithm for finding the greatest common divisor of two integers a and b.

6.1. **Definition.** The greatest common divisor of two integers a and b (not both zero) is the largest positive integer d which divides both $|a|$ and $|b|$.

The greatest common divisor satisfies the following conditions:

$$(6.2) \quad \begin{cases} \gcd(0, 0) \text{ is undefined} \\ \gcd(0, b) = |b| \qquad \text{for } b \neq 0 \\ \gcd(a, b) = \gcd(b, a) \\ \gcd(a, b) = \gcd(|a|, |b|). \end{cases}$$

Hence, there is no loss of generality if we assume that a and b are non-negative integers, not both zero.

The standard method for describing the Euclidean Algorithm for two unequal positive integers $a > b > 0$ is to use the *division property* that there exist integers q and r satisfying $q > 0$ and $0 \leq r < b$, such that

$$(6.3) \qquad a = bq + r.$$

Here q is the quotient and r is the remainder when a is divided by b. If we use this property repeatedly, we obtain the system of equations and inequalities

$$(6.4) \quad \begin{cases} a = bq_1 + r_1 & 0 < r_1 < b \\ b = r_1 q_2 + r_2 & 0 < r_2 < r_1 \\ r_1 = r_2 q_3 + r_3 & 0 < r_3 < r_2 \\ \quad \vdots & \quad \vdots \\ r_{n-2} = r_{n-1} q_n + r_n & 0 < r_n < r_{n-1} \\ r_{n-1} = r_n q_{n+1} \end{cases}$$

where $r_n \neq 0$ (but $r_{n+1} = 0$). This process is called the Euclidean Algorithm and r_n, the last non-zero remainder in (6.4), is known to have the following property.

6.5. Theorem.

$$r_n = \gcd(a, b).$$

PROOF.

$$\begin{aligned} \gcd(a, b) &= \gcd(b, r_1) \\ &= \gcd(r_1, r_2) \\ &\quad \vdots \\ &= \gcd(r_{n-1}, r_n) \\ &= \gcd(r_n, 0) \\ &= r_n. \end{aligned} \qquad \square$$

The next theorem expresses an important characterization of the greatest common divisor.

6.6. Theorem. *If $a > b > 0$ are two integers, then $\gcd(a, b)$ is the smallest positive integer d such that*

$$d = ax + by,$$

where $x, y \in \mathbb{I}$.

PROOF. See, for example, Pettofrezzo and Byrkit [1970], p. 34. $\qquad \square$

Observe that the integers x and y are not unique in this theorem because, for any $t \in \mathbb{I}$,

$$(6.7) \qquad d = a(x + bt) + b(y - at).$$

Observe, also, that not every linear combination of a and b yields d because, if

$$(6.8) \qquad d = ax + by,$$

then

(6.9)
$$(kd) = a(kx) + b(ky),$$

for any $k \in \mathbb{I}$.

To find a pair of integers x and y which satisfy (6.8), we use (6.4) and solve for the successive remainders $r_1, r_2, \ldots, r_{n+1}$ as functions of a and b. Hence,

(6.10)
$$\begin{cases}
r_1 = a + b(-q_1) \\
r_2 = b + r_1(-q_2) \\
\quad = a(-q_2) + b(1 + q_1 q_2) \\
r_3 = r_1 + r_2(-q_3) \\
\quad = a(1 + q_2 q_3) + b(-q_1 - q_3 - q_1 q_2 q_3) \\
\quad \vdots \\
r_n = r_{n-2} + r_{n-1}(-q_n) \\
\quad = ax + by \\
0 = r_{n-1} + r_n(-q_{n+1}) \\
\quad = au + bv.
\end{cases}$$

The computation indicated in (6.10) can be expressed in the form of a table.

6.11. Table. The computation of r_n, x, and y.

	a	1	0
	b	0	1
q_1	r_1	1	$-q_1$
q_2	r_2	$-q_2$	$1 + q_1 q_2$
q_3	r_3	$1 + q_2 q_3$	$-q_1 - q_3 - q_1 q_2 q_3$
$\vdots$	$\vdots$	$\vdots$	$\vdots$
q_n	r_n	x	y
q_{n+1}	0	u	v

Notice that the column (in Table 6.11) which contains the integers a and b displays the sequence of remainders generated by (6.4), the Euclidean Algorithm. When we add one or more columns (for example, the columns which generate the integers x and y) we have what Knuth [1981], p. 325, calls an *Extended Euclidean Algorithm*. See Algorithm 6.26, for example.

In Table 6.11 we are generating a sequence of $n + 1$ integer triples as follows, where the *seed matrix* is

$$\begin{bmatrix} a & 1 & 0 \\ b & 0 & 1 \end{bmatrix},$$

and the integer triples are written as row vectors. Thus,

$$[r_1, 1, -q_1] = [1, -q_1]\begin{bmatrix} a & 1 & 0 \\ b & 0 & 1 \end{bmatrix},$$

(6.12) $$[r_2, -q_2, 1 + q_1 q_2] = [1, -q_2]\begin{bmatrix} b & 0 & 1 \\ r_1 & 1 & -q_1 \end{bmatrix},$$

$$[r_3, 1 + q_2 q_3, -q_1 - q_3 - q_1 q_2 q_3] = [1, -q_3]\begin{bmatrix} r_1 & 1 & -q_1 \\ r_2 & -q_2 & 1 + q_1 q_2 \end{bmatrix},$$

and so on. In these equations $q_1, q_2, \ldots, q_{n+1}$ are given by*

(6.13)
$$\begin{cases} q_1 = \left[\dfrac{a}{b}\right] \\[2mm] q_2 = \left[\dfrac{b}{r_1}\right] \\[2mm] q_3 = \left[\dfrac{r_1}{r_2}\right] \\[1mm] \vdots \\[1mm] q_{n+1} = \left[\dfrac{r_{n-1}}{r_n}\right]. \end{cases}$$

6.14. EXAMPLE. To find $\gcd(19, 11)$ along with x and y (see Table 6.11), we use the seed matrix

$$\begin{bmatrix} 19 & 1 & 0 \\ 11 & 0 & 1 \end{bmatrix}.$$

The following table contains the computation.

	19	1	0
	11	0	1
1	8	1	−1
1	3	−1	2
2	2	3	−5
1	1	−4	7
2	0	11	−19

From this table we obtain

$$\gcd(19, 11) = 1$$

$$x = -4$$

$$y = 7.$$

*The symbol $[a/b]$ denotes the integer part of the quotient a/b.

As a check, observe that

$$1 = (19)(-4) + (11)(7).$$

Also,* since $u = 11$ and $v = -19$,

$$0 = (19)(11) + (11)(-19).$$

The multiplicative inverse modulo m

We find an application for the computation indicated in (6.10) and Table 6.11 when we wish to compute the multiplicative inverse of an integer b in the finite commutative ring $(\mathbb{I}_m, +, \cdot)$ or the finite field $(\mathbb{I}_p, +, \cdot)$ if $m = p$, a prime. From Corollary 2.17 we know that $b^{-1}(m)$ exists, for $b \neq 0$, if and only if $\gcd(m, b) = 1$. Therefore, the following theorem is of some importance.

6.15. Theorem. *If* $\gcd(m, b) = 1$ *and if*

$$1 = mx + by,$$

then

$$b^{-1}(m) = |y|_m.$$

PROOF. Since $1 = mx + by$ we can write

$$1 = |mx + by|_m$$
$$= |by|_m$$
$$= |b|y|_m|_m$$

and it follows that

$$|y|_m = b^{-1}(m),$$

by definition. $\qquad\qquad\square$

Therefore, in Example 6.14 we find that $y = 7$ implies $11^{-1}(19) = |7|_{19} = 7$. As a check, observe that $|11 \cdot 7|_{19} = 1$.

6.16. EXAMPLE. Suppose we choose $m = 5^4 = 625$ and we wish to compute the multiplicative inverse of 342 modulo 625. Since $\gcd(342, 5) = 1$, we know that this inverse exists. In this case the seed matrix is

$$\begin{bmatrix} 625 & 0 \\ 342 & 1 \end{bmatrix}$$

---

*See Problem 2, Exercises I.6, for a generalization.

and we obtain the table

	625	0
	342	1
1	283	-1
1	59	2
4	47	-9
1	12	11
3	11	-42
1	1	53
11	0	-625

Consequently, $\gcd(625, 342) = 1$ and $y = 53$, which implies

$$342^{-1} = |53|_{625}$$
$$= 53.$$

As a check we observe that

$$|53 \cdot 342|_{625} = 1.$$

The forward mapping

It is possible to implement the forward mapping $\mathbb{F}_N \to \hat{\mathbb{I}}_m$ by a two-step procedure. Thus, to find $|d/c|_m$ we first compute $c^{-1}(m)$ using the seed matrix

$$\begin{bmatrix} m & 0 \\ c & 1 \end{bmatrix}$$

as in Example 6.16. Next, we use Definition 5.1 to write $|d/c|_m = |d \cdot c^{-1}|_m$.

On the other hand, these two steps can be combined into a single step if we begin with the seed matrix

$$\begin{bmatrix} m & 0 \\ c & d \end{bmatrix}.$$

This was called to our attention by P. Kornerup. See Kornerup and Gregory [1983].

Observe that if $d = 1$, this reduces to the problem of computing

$$(6.17) \qquad\qquad \left|\frac{1}{c}\right|_m = c^{-1}(m).$$

In the next example we illustrate both the two-step method and the single-step method.

6.18. EXAMPLE. Suppose we wish to compute $|-\tfrac{3}{2}|_{19}$. For the two-step method we use the seed matrix

$$\begin{bmatrix} 19 & 0 \\ 2 & 1 \end{bmatrix}$$

and obtain the table

	19	0
	2	1
9	1	−9
2	0	19

Hence,

$$2^{-1} = |-9|_{19}$$
$$= 10.$$

Consequently,

$$|-\tfrac{3}{2}|_{19} = |(-3)(2^{-1})|_{19}$$
$$= |-30|_{19}$$
$$= 8.$$

For Kornerup's single-step method we use the seed matrix

$$\begin{bmatrix} 19 & 0 \\ 2 & -3 \end{bmatrix}$$

and obtain the table

	19	0
	2	−3
9	1	27
2	0	−57

Hence, since each entry in the third column has been multiplied by -3.

$$|-\tfrac{3}{2}|_{19} = |27|_{19}$$
$$= 8.$$

Some properties of an Extended Euclidean Algorithm

Suppose we select four intergers a, b, c, and d, and an arbitrary sequence of integers $\{q_1, q_2 \ldots\}$, and generate a sequence of integer pairs by the recursion

(6.19a)
$$\begin{cases} a_i = a_{i-2} - q_i a_{i-1} \\ b_i = b_{i-2} - q_i b_{i-1} \end{cases} \quad i = 1, 2, \dots.$$

In matrix language this becomes

(6.19b)
$$[a_i, b_i] = [1, -q_i] \begin{bmatrix} a_{i-2} & b_{i-2} \\ a_{i-1} & b_{i-1} \end{bmatrix},$$

where the seed matrix is

(6.20)
$$\begin{bmatrix} a_{-1} & b_{-1} \\ a_0 & b_0 \end{bmatrix} = \begin{bmatrix} a & b \\ c & d \end{bmatrix}.$$

We represent this computation in the following table:

	a	b
	c	d
q_1	a_1	b_1
q_2	a_2	b_2
q_3	a_3	b_3
.	.	.
.	.	.
.	.	.

6.21. EXAMPLE. Suppose $a = -3$, $b = -2$, $c = -2$, and $d = 5$, with $q_1 = 2$, $q_2 = -1$, $q_3 = 4$, ... The computation becomes

	-3	-2
	-2	5
2	1	-12
-1	-1	-7
4	5	16
.	.	.
.	.	.
.	.	.

If we continue to choose the sequence $\{q_1, q_2, \dots\}$ arbitrarily but impose (on the seed matrix) the condition

(6.22)
$$\begin{vmatrix} a & b \\ c & d \end{vmatrix} \equiv 0 \quad (\text{mod } w)$$

for some w, then we get the following result.

6.23. **Lemma.** *If $ad - bc \equiv 0$ (mod w), then for $i = 1, 2, \dots$*

$$a_i b_{i-1} - a_{i-1} b_i \equiv 0 \quad (\text{mod } w).$$

PROOF.

$$a_i b_{i-1} - a_{i-1} b_i = (a_{i-2} - q_i a_{i-1}) b_{i-1} - a_{i-1}(b_{i-2} - q_i b_{i-1})$$

$$= (a_{i-2} b_{i-1} - a_{i-1} b_{i-2}) + q_i \cdot 0$$

$$\vdots$$

$$= (-1)^{i-1}(a_{-1} b_0 - a_0 b_{-1})$$

$$= (-1)^{i-1}(ab - cd)$$

$$\equiv 0 \pmod{w}. \qquad \square$$

Observe that in Example 6.21 we have

(6.24)
$$\begin{vmatrix} -3 & -2 \\ -2 & 5 \end{vmatrix} \equiv \begin{vmatrix} -2 & 5 \\ 1 & -12 \end{vmatrix} \equiv \begin{vmatrix} 1 & -12 \\ -1 & -7 \end{vmatrix} \equiv \begin{vmatrix} -1 & -7 \\ 5 & 16 \end{vmatrix}$$
$$\equiv 0 \pmod{19},$$

which illustrates the lemma.

In other words, if the modulus is 19, the sequence of rational numbers corresponding to the integer pairs in Example 6.21, that is, the sequence

$$\{\tfrac{2}{3}, -\tfrac{5}{2}, -12, 7, \tfrac{16}{5}, \ldots\}$$

contains rational numbers, all of which belong to the same generalized residue class $\mathbb{Q}_7$. (See Corollary 5.7.)

Now suppose a and c are positive integers and the sequence $\{q_1, q_2, \ldots\}$ is chosen using the definition

(6.25)
$$q_i = \left[\frac{a_{i-2}}{a_{i-1}} \right],$$

for $a_{i-1} \neq 0$, as in (6.13). In this case, it is easy to verify that the (finite) sequence $\{a_1, a_2, \ldots, a_n, a_{n+1}\}$ is the sequence of partial remainders generated when the Euclidean Algorithm is applied to the problem of finding $\gcd(a, c)$. (See Table 6.11.) Therefore, $a_n = \gcd(a, c)$ and $a_{n+1} = 0$.

6.26. **Algorithm** (Extended Euclidean Algorithm).

(i) *Choose the seed matrix*

$$\begin{bmatrix} a_{-1} & b_{-1} \\ a_0 & b_0 \end{bmatrix} = \begin{bmatrix} a & b \\ c & d \end{bmatrix},$$

where a and c are positive integers and b and d are arbitrary integers.
(ii) *For $i = 1, 2, \ldots, n+1$, while $a_{i-1} \neq 0$, determine q_i as the quotient and a_i as the non-negative remainder in the division of a_{i-2} by a_{i-1}. Then*

$$a_i = a_{i-2} - q_i a_{i-1}.$$

(iii) *Likewise, define*

$$b_i = b_{i-2} - q_i b_{i-1}.$$

(iv) *Terminate when* $a_{n+1} = 0$. *At this point*

$$a_n = \gcd(a, c).$$

This extended algorithm can be used to carry out the forward mapping $|r/s|_m = |rs^{-1}|_m$ as we demonstrated in Example 6.18. We sometimes call the integer $|r/s|_m$ the least non-negative residue of r/s modulo m. For this application we need another lemma.

6.27. Lemma. *In Algorithm* 6.26 *choose* $a = m$, $b = 0$, *and* $0 < c < m$ *such that* $\gcd(c, m) = 1$. *Then, for* $i = 1, 2, \ldots, n$,

$$\left|\frac{b_i}{a_i}\right|_m = \left|\frac{d}{c}\right|_m$$

and alternatively, if $0 < |d| < m$ *such that* $\gcd(d, m) = 1$,

$$\left|\frac{a_i}{b_i}\right|_m = \left|\frac{c}{d}\right|_m.$$

PROOF. Since $a = m$ and $b = 0$, we have

$$ad \equiv bc \pmod{m}.$$

Hence, by Lemma 6.23,

$$a_1 d \equiv b_1 c \pmod{m}.$$

Consequently, since

$$\left|\frac{u}{v}\right|_m = \left|\frac{r}{s}\right|_m$$

if and only if

$$us \equiv vr \pmod{m},$$

the conclusion follows. $\square$

6.28. EXAMPLE. Let $m = 19$ and $d/c = \frac{8}{10}$. Then the seed matrix is

$$\begin{bmatrix} 19 & 0 \\ 10 & 8 \end{bmatrix}$$

and the computation is given in the following table:

	19	0
	10	8
1	9	-8
1	1	16
9	0	-152

Hence,

$$\left|\tfrac{8}{10}\right|_{19} = \left|\tfrac{-8}{9}\right|_{19} = |16|_{19} = 16.$$

6.29. Remark. We can't overemphasize the remarkable property of the recursion (6.19) that if

$$\begin{vmatrix} a & b \\ c & d \end{vmatrix} \equiv 0 \qquad (\mathrm{mod}\ w),$$

then all integer pairs generated represent rational numbers in the same generalized residue class $\mathbb{Q}_k$ for some k in the range $0 \le k < w$.

We now come to the main result.

6.30. Theorem. *Given any rational number r/s, with $s > 0$, and an integer m such that $\gcd(s, m) = 1$, Algorithm 6.26 seeded with the matrix*

$$\begin{bmatrix} m & 0 \\ s & r \end{bmatrix}$$

will terminate (for some $n + 1$ such that $a_{n+1} = 0$). At this point

$$\left|\frac{r}{s}\right|_m = |b_n|_m.$$

PROOF. Since $a_n = \gcd(s, m) = 1$, it follows from Lemma 6.27 that

$$\left|\frac{b_n}{a_n}\right|_m = |b_n|_m = \left|\frac{r}{s}\right|_m. \qquad \square$$

6.31. EXAMPLE. If we wish to compute $\left|\tfrac{10}{13}\right|_{625}$, we use the seed matrix

$$\begin{bmatrix} 625 & 0 \\ 13 & 10 \end{bmatrix}$$

in Algorithm 6.26. The computation appears in the following table:

		625	0
		13	10
48		1	-480
13		0	6250

from which we conclude that

$$\left|\tfrac{10}{13}\right|_{625} = |-480|_{625}$$
$$= 145.$$

We point out that, in this example, $m = 5^4 = 625$ is not a prime but $\gcd(625, 13) = 1$ and Theorem 6.30 applies.

In Example 6.31 the fraction $\frac{10}{13}$ is an order-17 Farey fraction. However, it is not necessary that the fraction r/s in Theorem 6.30 be an order-N Farey fraction. For example, if we choose $r/s = -\frac{5}{56}$ and apply the algorithm, we find that it, too, is mapped onto the integer 145 in $\hat{\mathbb{I}}_{625}$. Therefore, both $\frac{10}{13}$ and $-\frac{5}{56}$ belong to the same generalized residue class $\mathbb{Q}_{145}$. See Example 6.33.

The inverse mapping (Kornerup and Krishnamurthy)

With (6.19) we can generate an infinite sequence of integer pairs $\{(a_1, b_1), (a_2, b_2), \ldots\}$. Lemmas 6.23 and 6.27 taken together (as in Theorem 6.30) state that with the seed matrix

$$\begin{bmatrix} m & 0 \\ s & r \end{bmatrix}$$

and *any* sequence $\{q_1, q_2, \ldots\}$, the infinite sequence of rational numbers associated with the integer pairs generated, that is, the sequence

$$\left\{ \frac{b_1}{a_1}, \frac{b_2}{a_2}, \ldots \right\}$$

has the property that for $i = 1, 2, \ldots$

(6.32)
$$\left| \frac{b_i}{a_i} \right|_m = \left| \frac{r}{s} \right|_m.$$

Therefore, it is possible to generate an infinite number of fractions from the same generalized residue class $\mathbb{Q}_k$, where $0 \le k < m$, by choosing $(s, r) = (k, 1)$. Thus, we have a method for implementing the inverse mapping $\hat{\mathbb{I}}_m \to \mathbb{F}_N$, if we can select among the elements of $\mathbb{Q}_k$ the (unique) order-N Farey fraction.

6.33. EXAMPLE. We shall "invert" the mapping exhibited in the previous example, that is, we shall map the integer $145 \in \hat{\mathbb{I}}_{625}$ onto $\frac{10}{13} \in \mathbb{F}_{17}$.
     Since $k = 145$, our seed matrix is

$$\begin{bmatrix} 625 & 0 \\ 145 & 1 \end{bmatrix}.$$

The computation is displayed in the following table, where each q_i is chosen using (6.25). With this choice, the table has finite length.

	625	0
	145	1
4	45	−4
3	10	13
4	5	−56
2	0	125

Notice that $\frac{10}{13}$ is recovered. Notice, also, that 145 and 625 are not relatively prime and that

$$a_n = \gcd(625, 145)$$

$$= 5,$$

as described in Algorithm 6.26.

It is easy to recover the order-17 Farey fraction $\frac{10}{13}$ from the table above, because it is the only element of the set

$$\{-\tfrac{45}{4}, \tfrac{10}{13}, -\tfrac{5}{56}\} \subset \mathbb{Q}_{145}$$

which is also an element of $\mathbb{F}_{17}$.

Recall that when $\mathbb{Q}_k$ contains an order-N Farey fraction, its uniqueness is guaranteed by Theorem 5.14. The obvious question to raise here is whether or not we can guarantee that this unique order-N Farey fraction will always appear in the finite subset of $\mathbb{Q}_k$ generated in the table above. It turns out that when we use Algorithm 6.26, which uses the integers $\{q_1, q_2, \ldots, q_{n+1}\}$ defined in (6.25), we can indeed guarantee the recovery of the order-N Farey fraction belonging to $\mathbb{Q}_k$, that is, we can carry out the inverse mapping. To show this, we need to prove two additional theorems. See Kornerup and Gregory [1983].

Let us extend the seed matrix by writing

$$(6.34) \qquad \begin{bmatrix} a_{-1} & b_{-1} & c_{-1} \\ a_0 & b_0 & c_0 \end{bmatrix} = \begin{bmatrix} m & 0 & -1 \\ k & 1 & 0 \end{bmatrix},$$

and let us define the sequence of integer triples $\{(a_i, b_i, c_i)\}$ by the recursion, while $a_{i-1} \neq 0$,

$$(6.35) \qquad q_i = \left\lceil \frac{a_{i-2}}{a_{i-1}} \right\rceil \quad \text{and} \quad \begin{cases} a_i = a_{i-2} - q_i a_{i-1} \\ b_i = b_{i-2} - q_i b_{i-1} \\ c_i = c_{i-2} - q_i c_{i-1}, \end{cases}$$

for $i = 1, 2, \ldots, n + 1$. It is well known (see Hardy and Wright [1960], for example) that the sequence

(6.36)
$$\left\{ \frac{|b_1|}{|c_1|}, \frac{|b_2|}{|c_2|}, \ldots, \frac{|b_{n+1}|}{|c_{n+1}|} \right\}$$

is the complete sequence of continued fraction convergents of m/k. It can also be shown* that, for $i = 1, 2, \ldots, n + 1$.

(6.37)
$$a_i = kb_i - mc_i.$$

These continued fraction convergents are the so-called "best rational approximations", for which the following theorem holds.

6.38. **Theorem.** *Every fraction r/s which satisfies the inequality*

$$\left| \alpha - \frac{r}{s} \right| < \frac{1}{2s^2}$$

is a continued fraction convergent of α.

PROOF. See Hardy and Wright [1960], p. 153.   □

We may now prove the following theorem which states that if an order-N Farey fraction belongs to the generalized residue class $\mathbb{Q}_k$, it will always be recovered by Algorithm 6.26.

6.39. **Theorem** (Kornerup). *If r/s is an order-N Farey fraction satisfying*

$$\left| \frac{r}{s} \right|_m = |k|_m = k$$

where $k \in \hat{\mathbb{I}}_m$ and

$$0 < r \leq N$$

$$0 < |s| \leq N,$$

then there exists an integer i such that

$$(r, s) = (a_i, b_i),$$

where $\{(a_j, b_j)\}, j = 1, 2, \ldots, n + 1$, is the sequence of integer pairs generated by Algorithm 6.26 seeded with the matrix

$$\begin{bmatrix} m & 0 \\ k & 1 \end{bmatrix}.$$

PROOF. If we extend the seed matrix as in (6.34) and define the sequence $\{c_i\}$, $i = 1, 2, \ldots, n + 1$, as in (6.35), then (6.36) is the complete sequence of continued fraction convergents of m/k whenever $k \neq 0$. From our hypothesis

---
*See Problem 2 in Exercises I.6.

$$\left|\frac{r}{s}\right|_m = |k|_m = k \in \hat{\mathbb{I}}_m.$$

Therefore,

$$r \equiv ks \quad (\mathrm{mod}\ m),$$

which implies the existence of a unique integer t such that

$$r = ks - mt.$$

This allows us to write

$$\left|\frac{k}{m} - \frac{t}{s}\right| = \left|\frac{ks - mt}{ms}\right|$$

$$= \left|\frac{r}{ms}\right|$$

$$\leq \frac{1}{s^2} \cdot \frac{|s| \cdot N}{2N^2 + 1}$$

$$\leq \frac{1}{s^2} \cdot \frac{N^2}{2N^2 + 1}$$

$$< \frac{1}{2s^2}.$$

Thus, using Theorem 6.38, we deduce that either t/s or $-t/-s$ is a continued fraction convergent of k/m.

Since (6.36) is the sequence of continued fraction convergents of m/k, it follows that

$$\left\{\frac{0}{1}, \frac{|c_1|}{|b_1|}, \ldots, \frac{|c_{n+1}|}{|b_{n+1}|}\right\}$$

is the sequence of continued fraction convergents of k/m. Hence, there exists an integer i, where $1 \leq i \leq n + 1$ such that $|b_i| = |s|$ and

$$\frac{t}{s} = \frac{c_i}{b_i}.$$

Finally, from (6.37), we have

$$\frac{a_i}{b_i} = k - m\frac{c_i}{b_i}$$

and from $r = ks - mt$, we have

$$\frac{r}{s} = k - m\frac{t}{s}.$$

Hence,

$$\frac{a_i}{b_i} = \frac{r}{s},$$

and so

$$(r, s) = (a_i, b_i),$$

since both r and a_i are positive.   □

The "common denominator" method

If the order-N Farey fraction a/b is the image of the integer k in $\hat{\mathbb{I}}_m$, and if a multiple of b can be found, then there is a simpler method than Algorithm 6.26 for mapping k onto a/b.

6.40. Theorem. *Suppose $x = a/b$ is an order-N Farey fraction and*

$$k = |ab^{-1}|_m.$$

We can map k onto x as follows: if tb can be found with the integer t in the range $0 < t \le N$, then

$$ta = |(tb)k|_m,$$

where $|\cdot|_m$ is defined in Definition 2.24. Obviously,

$$x = \frac{ta}{tb}.$$

PROOF. From the hypothesis we can write $ta = (tb)x$, which implies

$$|ta|_m = |(tb)x|_m$$
$$= |(tb)|x|_m|_m$$
$$= |(tb)k|_m.$$

However, since $|a| \le N$ and $0 < t \le N$,

$$|ta| \le N^2$$
$$\le (m - 1)/2.$$

Therefore, $|ta|_m = ta$, which implies

$$ta = |(tb)k|_m$$

and

$$x = \frac{ta}{tb}.$$ □

6.41. EXAMPLE. Let us recall the example in Remark 5.23,

$$x = \tfrac{1}{2} + (-\tfrac{2}{3}) + (-\tfrac{1}{6}).$$

In $(\mathbb{I}_{19}, +, \cdot)$ we obtain

$$
\begin{aligned}
|x|_{19} &= |\tfrac{1}{2} + (-\tfrac{2}{3}) + (-\tfrac{1}{6})|_{19} \\
&= |10 + 12 + 3|_{19} \\
&= 6.
\end{aligned}
$$

Consequently, $k = 6$. Also, we observe that a common denominator for the three rational numbers in the computation is 6. Hence, we can use $tb = 6$ and obtain

$$
\begin{aligned}
ta &= /(tb)k/_{19} \\
&= /6 \cdot 6/_{19} \\
&= -2.
\end{aligned}
$$

Therefore,

$$
\begin{aligned}
x &= \frac{ta}{tb} \\
&= -\tfrac{1}{3}.
\end{aligned}
$$

In this computation $m = 19$, $N = 3$, and $t = 2$. Thus, the inequality $0 < t \le N$ is satisfied and we obtain the correct result. No method is known, a priori, for finding an integer t in the range $0 < t \le N$. What we can do, however, is use *any* multiple of b, and then examine the result to see whether or not it is an element of $\mathbb{F}_N$.

6.42. EXAMPLE. Suppose $m = 19$, $N = 3$, $k = |x|_{19} = 8$, and the element we seek in $\mathbb{F}_3$ is

$$
x = -\tfrac{3}{2}.
$$

We can make use of the fact that $b = 2$ (this is usually not known, of course) and demonstrate what happens if we pick $tb = 6$ and $tb = 8$.

For $tb = 6$, we obtain

$$
\begin{aligned}
ta &= /6 \cdot 8/_{19} \\
&= -9,
\end{aligned}
$$

which implies

$$
x = -\tfrac{3}{2}.
$$

Since $x \in \mathbb{F}_3$, we have the correct answer. (Here, $t = 3$.)

For $tb = 8$, we obtain

$$
\begin{aligned}
ta &= /8 \cdot 8/_{19} \\
&= 7,
\end{aligned}
$$

which implies

$$x = \tfrac{7}{8}.$$

Since both numerator and denominator exceed $N = 3$, we have pseudo-overflow. This is not surprising, since $tb = 8$ implies $t = 4$, and $4 > N$.

Observe, however, that both $-\tfrac{3}{2}$ and $\tfrac{7}{8}$ belong to the same generalized residue class, because

$$\left|-\tfrac{3}{2}\right|_{19} = \left|\tfrac{7}{8}\right|_{19} = 8.$$

Obviously, $-\tfrac{3}{2}$ is the unique order-3 Farey fraction in $\mathbb{Q}_8$ and $\tfrac{7}{8}$ is another one of the infinitely many elements in $\mathbb{Q}_8$ which satisfy

$$\left|\frac{a}{b}\right|_{19} = 8.$$

6.43. Remark. The requirement that t satisfy the inequality

$$0 < t \le N$$

in Theorem 6.40 is *sufficient* but not *necessary*. To see this, we return to Example 6.41 in which

$$|x|_{19} = 6.$$

If we use $tb = 27$, we obtain

$$ta = /27 \cdot 6/_{19}$$
$$= -9,$$

which implies

$$x = -\tfrac{1}{3},$$

and this is the value we obtained earlier. In this case $t = 9$ which is larger than $N = 3$.

EXERCISES I.6

1. Let $a = 49$ and $b = 63$. Find integers x and y such that $\gcd(a, b) = ax + by$.

2. We write Table 6.11 using the following notation:

	a	1	0
	b	0	1
q_1	r_1	s_1	t_1
$\vdots$	$\vdots$	$\vdots$	$\vdots$
q_n	r_n	s_n	t_n
q_{n+1}	0	s_{n+1}	t_{n+1}

Prove that $r_i = as_i + bt_i$, for $i = 1, 2, \ldots, n + 1$.

3. Compute the multiplicative inverse of 321 modulo 625.

4. Compute $\left|-\frac{15}{14}\right|_{625}$ and $\left|-\frac{4}{5}\right|_{81}$.

5. If $m = 625$, $N = 17$, which order-N Farey fractions correspond to the integers 259, 367, and 530? (Use the inverse mapping described in this section.)

6. If $m = 81$, $N = 6$, which order-N Farey fractions correspond to the integers 16, 39, and 66? See Problem 5.

7. Use $m = 625$ and single-modulus residue arithmetic to evaluate

$$x = \frac{\frac{3}{7} - \frac{5}{11}}{\frac{7}{11} - \frac{3}{11}}.$$

8. Prove that even though 369 is an element of $\mathbb{I}_{625}$, it is not an element of $\hat{\mathbb{I}}_{625}$.

7 Multiple-Modulus Residue Arithmetic with Rational Numbers

As in Section 3 we begin with a base vector β whose components are the moduli m_1, m_2, $\ldots$, m_n with $M = m_1 m_2 \ldots m_n$. Recall (3.1), (3.2) and (3.5). In this section we shall assume that N is the largest integer satisfying the inequality

(7.1)
$$2N^2 + 1 \le M.$$

We also assume that each modulus m_i is a prime which guarantees that for all $i, j = 1, 2, \ldots n$ and $i \ne j$,

(7.2)
$$\gcd(m_i, m_j) = 1.$$

In Section 5 we showed that the system $(\hat{\mathbb{Q}}, +, \cdot)$ is a commutative ring with identity (Theorem 5.10) and that a homomorphism exists between that ring and the finite commutative ring $(\mathbb{I}_m, +, \cdot)$, where $m = p^r$ for some prime p. Since multiple-modulus residue arithmetic using the base vector β is equivalent to single-modulus residue arithmetic with modulus M, we need to establish an analogous homomorphism between a similar commutative ring with identity, $(\tilde{\mathbb{Q}}, +, \cdot)$, and the finite commutative ring $(\mathbb{I}_M, +, \cdot)$.

The development completely parallels the development in Section 5 with a few changes reflecting the fact that, in Section 5, $m = p^r$ is a power of a single prime, whereas in this section, $M = m_1 m_2 \ldots m_n$ is the product of n distinct primes.

To map a rational number $x = a/b$ onto an integer in $\mathbb{I}_M$ we must be able to compute $b^{-1}(M)$ so that, as in Definition 5.1,

(7.3)
$$\left|\frac{a}{b}\right|_M = |ab^{-1}|_M.$$

50 I Residue or Modular Arithmetic

In this case $b^{-1}(M)$ exists if and only if $\gcd(b, M) = 1$, and this is true if and only if

(7.4)
$$\gcd(b, m_i) = 1,$$

for $i = 1, 2, \ldots, n$. Consequently, the only rational numbers which cannot be mapped into $\mathbb{I}_M$ using (7.3) are those rational numbers for which (7.4) fails for at least one value of i.

Let $\tilde{\mathbb{Q}}$ be the set of rational numbers which can be mapped onto $\mathbb{I}_M$ by (7.3), that is, let

(7.5)
$$\tilde{\mathbb{Q}} = \left\{ \frac{a}{b} : \gcd(b, m_i) = 1, \text{ for } i = 1, 2, \ldots, n \right\}.$$

Then (7.3) describes the mapping $|\cdot|_M : \tilde{\mathbb{Q}} \to \mathbb{I}_M$.

It turns out that each integer $j \in \mathbb{I}_M$ is the image of an infinite set of elements in $\tilde{\mathbb{Q}}$ which we shall label $\mathbb{Q}_j$. Thus, for $j = 0, 1, \ldots, M - 1$,

(7.6)
$$\mathbb{Q}_j = \left\{ \frac{a}{b} \in \tilde{\mathbb{Q}} : \left| \frac{a}{b} \right|_M = j \right\}$$

from which it follows that

(7.7)
$$\tilde{\mathbb{Q}} = \bigcup_{j=0}^{M-1} \mathbb{Q}_j.$$

The set $\mathbb{Q}_0$ (the rational numbers in $\tilde{\mathbb{Q}}$ which are mapped onto zero) consists of those rational numbers a/b, with $\gcd(b, m_i) = 1$, for $i = 1, 2, \ldots, n$, for which $|a|_M = 0$. As in Section 5, we call the disjoint subsets $\mathbb{Q}_0, \mathbb{Q}_1, \ldots, \mathbb{Q}_{M-1}$, generalized residue classes modulo M. Analogous to Theorem 5.6 and Corollary 5.7 we have the following:

7.8. Theorem. Let $x = a/b$ and $y = c/d$, where both $b^{-1}(M)$ and $d^{-1}(M)$ exist. Then

$$|x|_M = |y|_M$$

if and only if

$$ad \equiv bc \pmod{M}.$$

PROOF. See the proof of Theorem 5.6. $\square$

7.9. Corollary. Let $x = a/b$ and $y = c/d$ be elements of $\tilde{\mathbb{Q}}$. Then x and y belong to the same generalized residue class $\mathbb{Q}_j$ if and only if

$$ad \equiv bc \pmod{M}.$$

PROOF. See the proof of Corollary 5.7. $\square$

7.10. Remark. Since $x = a/b$ is a rational number for which $|x|_M$ does *not* exist if and only if $\gcd(b, m_i) \neq 1$ for at least one i, and since we are assuming that $\gcd(a, b) = 1$, we conclude that $|x|_M$ fails to exist if and only if b is an integral multiple of m_i for at least one i.

Thus, establishing results analogous to Lemma 5.9, Theorem 5.10, and Lemma 5.11, we are led to the following result (which is analogous to Theorem 5.12).

7.11. Theorem. *The mapping* $|\cdot|_M : \tilde{\mathbb{Q}} \to \mathbb{I}_M$ *is a homomorphism with respect to addition and multiplication.*

PROOF. See the development of Theorem 5.12. □

Therefore, arithmetic operations in the ring $(\tilde{\mathbb{Q}}, +, \cdot)$ correspond to those same arithmetic operations in the ring $(\mathbb{I}_M, +, \cdot)$. Consequently, as in Section 5, we are interested in replacing arithmetic operations on certain rational numbers in $\tilde{\mathbb{Q}}$ by those same arithmetic operations on integers in $\mathbb{I}_M$.

It is not difficult to prove a theorem analogous to Theorem 5.14 showing that if the generalized residue class $\mathbb{Q}_j$ contains the order-N Farey fraction $x = a/b$, with N defined by (7.1), then x is the only order-N Farey fraction contained in $\mathbb{Q}_j$. Also, since the number of order-N Farey fractions is fewer than M, not every generalized residue class contains an element of $\mathbb{F}_N$.

If we denote the images of the order-N Farey fractions by

$$(7.12) \qquad \tilde{\mathbb{I}}_M = \left\{ \left| \frac{a}{b} \right|_M : \frac{a}{b} \in \mathbb{F}_N \right\},$$

then the mapping $|\cdot|_M : \mathbb{F}_N \to \tilde{\mathbb{I}}_M$ is one-to-one and onto and thus has an inverse. (Recall Theorem 5.17).

7.13. EXAMPLE. Let $m_1 = 5$ and $m_2 = 7$, so that $M = 35$. Then $N = 4$ and the mapping $|\cdot|_{35} : \mathbb{F}_4 \to \tilde{\mathbb{I}}_{35}$ is exhibited in the following table:

0	0		
1	1	-1	34
2	2	-2	33
3	3	-3	32
4	4	-4	31
.	.	.	.
$-\frac{3}{4}$	8	$\frac{3}{4}$	27
$\frac{1}{4}$	9	$-\frac{1}{4}$	26
.	.	.	.
$-\frac{2}{3}$	11	$\frac{2}{3}$	24
$\frac{1}{3}$	12	$-\frac{1}{3}$	23
$\frac{4}{3}$	13	$-\frac{4}{3}$	22
.	.	.	.
$-\frac{3}{2}$	16	$\frac{3}{2}$	19
$-\frac{1}{2}$	17	$\frac{1}{2}$	18

Obviously, $\tilde{\mathbb{I}}_{35} \subset \mathbb{I}_{35}$. Observe that the order-4 Farey fractions mapped onto $\tilde{\mathbb{I}}_{35}$ appear in clusters of at least two elements per cluster, and within each cluster the elements differ by unity. Also, observe the "skew symmetry" as we compare the two columns of the table. These properties are also exhibited in Example 5.18 and in the tables in Rao and Gregory [1981].

A two-modulus system

Obviously, what we have done up to this point is completely analogous to what we did in Section 5 for the single-modulus system, the only difference being the replacement of the modulus $m = p^r$ by the modulus $M = m_1 m_2 \ldots m_n$. We now consider the mapping of the order-N Farey fractions, not onto $\tilde{\mathbb{I}}_M$, but onto the smaller sets $\mathbb{I}_{m_1}, \mathbb{I}_{m_2}, \ldots, \mathbb{I}_{m_n}$.

7.14. EXAMPLE. Let us continue with $m_1 = 5$ and $m_2 = 7$ as in Example 7.13. When $m_1 = 5$ and $m_2 = 7$ we find that many of the order-4 Farey fractions are mapped onto the same integer. Thus, when $m_1 = 5$ we have

0	0					
1	$-\frac{1}{4}$	-4	1	$-\frac{2}{3}$	$-\frac{3}{2}$	
2	$-\frac{4}{3}$	$\frac{3}{4}$	-3	2	$\frac{1}{3}$	$-\frac{1}{2}$
3	$\frac{1}{2}$	$-\frac{1}{3}$	-2	3	$-\frac{3}{4}$	$\frac{4}{3}$
4	$\frac{3}{2}$	$\frac{2}{3}$	-1	4	$\frac{1}{4}$	

and when $m_2 = 7$ we have

0	0			
1	$-\frac{4}{3}$	1	$-\frac{3}{4}$	
2	$-\frac{1}{3}$	2	$\frac{1}{4}$	$-\frac{3}{2}$
3	$\frac{2}{3}$	-4	3	$-\frac{1}{2}$
4	$\frac{1}{2}$	-3	4	$-\frac{2}{3}$
5	$\frac{3}{2}$	$-\frac{1}{4}$	-2	$\frac{1}{3}$
6	$\frac{3}{4}$	-1	$\frac{4}{3}$	

In this example, both $m_1 = 5$ and $m_2 = 7$ are greater than $N = 4$. Hence, among the non-zero order-4 Farey fractions, none of the denominators fails to satisfy (7.4). Thus, the multiplicative inverses exist with respect to both moduli and so $|a/b|_{m_i}$ exists for $i = 1, 2$. An obvious question is whether or not we can always choose our moduli $m_1, m_2, \ldots, m_n$ to satisfy

(7.15)
$$m_i > N,$$

for $i = 1, 2, \ldots, n$. It turns out that the answer is yes for $n = 2$, but no for $n > 2$.

7.16. Theorem. *Let $M = m_1 m_2$, where m_1 and m_2 are successive primes, and let $N > 0$ be the largest integer which satisfies the inequality*

$$2N^2 + 1 \leq M.$$

Then m_1 and m_2 are both greater than N.

PROOF. (Matula). Let m_1 and m_2 be successive primes. Then,

$$m_1 < m_2 < 2m_1$$

from Bertrand's Postulate. (See Hardy and Wright [1960], p. 343.) Consequently,

$$2m_1^2 > m_1 m_2$$
$$\geq 2N^2 + 1,$$

from the hypothesis. But this implies $m_1 > N$. Thus, both m_1 and m_2 are greater than N. □

7.17. Theorem. *Let $m_1 < m_2 < \ldots < m_n$ be primes and let M be their product. Choose $N > 0$ to be the largest integer which satisfies the inequality*

$$2N^2 + 1 \leq M.$$

If $n > 2$, then $m_1 \leq N$.

PROOF. From the hypothesis

$$N \leq \left[\frac{(m_1 m_2 \ldots m_n) - 1}{2} \right]^{1/2} < N + 1,$$

and this implies

$$\frac{(m_1 m_2 \ldots m_n) - 1}{2} < (N + 1)^2.$$

Consequently,

$$m_1 m_2 \ldots m_n < 2N^2 + 4N + 3.$$

If $m_2 \leq N$, the theorem is proved. We now assume $m_2 > N$, which implies

$$m_1 < \frac{2N^2 + 4N + 3}{N^{n-1}}$$

$$= \frac{2}{N^{n-3}} + \frac{4}{N^{n-2}} + \frac{3}{N^{n-1}}.$$

Since the first three primes are 2, 3, and 5, $N \geq 3$. Thus, if $n > 2$, we obtain

$$m_1 < 2 + \tfrac{4}{3} + \tfrac{1}{3} = \tfrac{11}{3},$$

which implies $m_1 \leq 3 \leq N$.   □

7.18. **Remark.** When $n = 2$ we can always find two successive primes greater than N which satisfy the inequality

$$m_1 m_2 \geq 2N^2 + 1,$$

from Theorem 7.16. However, observe that it is not necessary that they be successive primes as we observe in the following table:

N	$2N^2 + 1$	$M = m_1 m_2$	An alternate choice
3	19		
4	33	$35 = 5 \cdot 7$	
5	51		
6	73	$77 = 7 \cdot 11$	$91 = 7 \cdot 13$
7	99		
8	129	$143 = 11 \cdot 13$	
9	163	$187 = 11 \cdot 17$	
10	201	$209 = 11 \cdot 19$	$221 = 13 \cdot 17$
11	243	$247 = 13 \cdot 19$	
12	289	$299 = 13 \cdot 23$	$323 = 17 \cdot 19$

7.19. PROBLEM. Suppose we take another look at Problem 5.20 and compute the simple result

$$x = \tfrac{1}{3} + \left(-\tfrac{2}{3}\right)$$

using the two-modulus residue system with base vector

$$\beta = [5, 7].$$

In $(\mathbb{I}_5, +, \cdot)$ we compute

$$|x|_5 = \left|\left|\tfrac{1}{3}\right|_5 + \left|-\tfrac{2}{3}\right|_5\right|_5$$
$$= |2 + 1|_5$$
$$= 3,$$

and in $(\mathbb{I}_7, +, \cdot)$ we compute

$$|x|_7 = \left|\left|\tfrac{1}{3}\right|_7 + \left|-\tfrac{2}{3}\right|_7\right|_7$$
$$= |5 + 4|_7$$
$$= 2.$$

This is equivalent to writing

$$\left|\tfrac{1}{3}\right|_\beta = [2, 5]$$

and

$$\left|-\tfrac{2}{3}\right|_\beta = [1, 4],$$

and forming the componentwise sum

$$|x|_\beta = [3, 2].$$

Now we use the method exhibited in Problem 4.39 to find the integer $|x|_{35} \in \tilde{\mathbb{I}}_{35}$ which corresponds to the solution $x \in \mathbb{F}_4$. The computation is presented in the following table:

β	$m_1 = 5$	$m_2 = 7$	
$|t_1|_\beta$	3	2	
$|d_0|_\beta$	3	3	subtract
$|t_1 - d_0|_\beta$	0	6	
$m_1^{-1}(\beta_1)$		3	multiply
$|t_2|_{\beta_1} = |(t_1 - d_0)/m_1|_{\beta_1}$		4	

Hence,

$$|x|_{35} = 3 + 4 \cdot 5$$
$$= 23,$$

and this integer in $\tilde{\mathbb{I}}_{35}$ is mapped onto $-\tfrac{1}{3}$ in $\mathbb{F}_4$ using the inverse mapping illustrated in Example 6.33, that is,

	35	0
	23	1
1	12	-1
1	11	2
1	1	-3
11	0	35

Observe that

$$x = -\tfrac{1}{3},$$

is the only order-4 Farey fraction among the fractions $-\tfrac{12}{1}, \tfrac{11}{2}$, and $-\tfrac{1}{3}$, all of which lie in $\mathbb{Q}_{23}$.

A multiple-modulus residue system with $n > 2$

The material beginning at this point, and continuing to the end of this section, is material which has been written since the book was sent to be typeset. Thus, none of this material is used elsewhere in the book; for example, it is not used in the "application chapters." However, despite this fact, we are adding this material (before the book goes to press) because of its importance.

From Theorem 7.16 and 7.17 it might be assumed that it is not practical to use more than two moduli in a multiple-modulus residue system with rational operands. This is not the case, however, since we will describe a practical method for handling three or more moduli. It was discovered by David Matula and modified by Carl Gregory (the first author's son). Because a rigorous treatment of the method will appear in a paper by Matula and Gregory (soon after this book is published) we do not include the derivation of the method here.

In order to describe the Matula–Gregory algorithm, we introduce the base vector

$$(7.20) \qquad \beta = [m_1, m_2, \ldots, m_n],$$

where $m_1, m_2, \ldots, m_n$ are distinct primes and M is their product. Consider any nonzero rational number a/b. We can write, for each modulus m_i,

$$(7.21) \qquad \frac{a}{b} = \frac{a_i}{b_i} \cdot (m_i)^{r_i},$$

where

$$(7.22) \qquad \gcd(a_i, b_i) = \gcd(a_i, m_i) = \gcd(b_i, m_i) = 1.$$

Obviously, the integer r_i can be positive negative, or zero.

Now let N be the largest integer such that

$$(7.23) \qquad 2N^2 + 1 \le M.$$

Our objective is to establish a bijective mapping between F_N and a set of n-tuples, and we proceed as follows.

7.24. Definition. For any nonzero rational number a/b,

$$\left| \frac{a}{b} \right|_\beta = \left[\left| \frac{a}{b} \right|^*_{m_1}, \left| \frac{a}{b} \right|^*_{m_2}, \ldots, \left| \frac{a}{b} \right|^*_{m_n} \right],$$

where, for $i = 1, 2, \ldots, n$, $\left| \dfrac{a}{b} \right|^*_{m_i}$ is the ordered pair

$$\left| \frac{a}{b} \right|^*_{m_i} = \left(\left| \frac{a_i}{b_i} \right|_{m_i}, r_i \right).$$

To represent zero, we may use any integer z (we usually choose $z = 0$) and write the ordered pairs

$$|0|^*_{m_i} = (0, z_i),$$

for $i = 1, 2, \ldots, n$.

7.25. EXAMPLE. Let $\beta = [2, 3, 5, 7]$ and let $a/b = 3/7$. Since

$$\left|\tfrac{3}{7}\right|^*_2 = (1, 0),$$
$$\left|\tfrac{3}{7}\right|^*_3 = (1, 1),$$
$$\left|\tfrac{3}{7}\right|^*_5 = (4, 0),$$
$$\left|\tfrac{3}{7}\right|^*_7 = (3, -1),$$

we have

$$\left|\tfrac{3}{7}\right|_\beta = [(1, 0), (1, 1), (4, 0), (3, -1)].$$

Observe that $r_i > 0$ indicates that m_i is a factor of the numerator, and $r_i < 0$ indicates that m_i is a factor of the denominator. If m_i divides neither the numerator nor the denominator, $r_i = 0$.

From (7.21) and Definition 7.24 it is clear that, for $i = 1, 2, \ldots, n$,

$$(7.26) \qquad \left|\frac{b}{a}\right|^*_{m_i} = \left(\left|\frac{b_i}{a_i}\right|_{m_i}, -r_i\right),$$

and this enables us to represent the multiplicative inverse of any nonzero rational number whose representation is known.

7.27. EXAMPLE. In Example 7.25 we have $\left|\tfrac{3}{7}\right|_\beta$. In order to obtain $\left|\tfrac{7}{3}\right|_\beta$ we compute the multiplicative inverse of the first component in each ordered pair and change the sign of the second component. Thus, since $\beta = [2, 3, 5, 7]$, and since

$$(1, 0) \to (1, 0),$$
$$(1, 1) \to (1, -1),$$
$$(4, 0) \to (4, 0),$$
$$(3, -1) \to (5, 1),$$

we have

$$\left|\tfrac{7}{3}\right|_\beta = [(1, 0), (1, -1), (4, 0), (5, 1)].$$

It is easy to show that, for $i = 1, 2, \ldots, n$,

$$(7.28) \qquad \left|-\frac{a}{b}\right|^*_{m_i} = \left(\left|-\frac{a_i}{b_i}\right|_{m_i}, r_i\right),$$

and this enables us to represent the additive inverse of any rational number whose representation is known.

7.29. EXAMPLE. In Example 7.25 we have $\left|\frac{3}{7}\right|_\beta$. In order to obtain $\left|-\frac{3}{7}\right|_\beta$ we compute the additive inverse of the first component in each ordered pair and leave the second component unchanged. Thus, since $\beta = [2, 3, 5, 7]$, and since

$$(1, 0) \to (1, 0),$$
$$(1, 1) \to (2, 1),$$
$$(4, 0) \to (1, 0),$$
$$(3, -1) \to (4, -1),$$

we have

$$\left|-\tfrac{3}{7}\right|_\beta = [(1, 0), (2, 1), (1, 0), (4, -1)].$$

Consider the set of images of the set of rational numbers

(7.30) $$\mathbb{T}_\beta = \left\{ \left|\frac{a}{b}\right|_\beta : \frac{a}{b} \in \mathbb{Q} \right\},$$

and the finite subset of $\mathbb{T}_\beta$ consisting of the images of the order-N Farey fractions, that is,

(7.31) $$\tilde{\mathbb{T}}_\beta = \left\{ \left|\frac{a}{b}\right|_\beta : \frac{a}{b} \in \mathbb{F}_N \right\}.$$

The mapping $|\cdot|_\beta : \mathbb{F}_N \to \tilde{\mathbb{T}}_\beta$ is onto (a surjection) by (7.31). We will show that it is one-to-one (an injection) by exhibiting the inverse mapping $\tilde{\mathbb{T}}_\beta \to \mathbb{F}_N$.

7.32. **Definition.** Let the symbols M_r, M_0, and M_{-r} be defined as follows, for each $|a/b|_\beta$ under consideration:

(i) $$M_r = \prod_i m_i$$

where m_i is a modulus for which $r_i > 0$ in Definition 7.24. If no such modulus exists, $M_r = 1$.

(ii) $$M_0 = \prod_j m_j$$

where m_j is a modulus for which $r_j = 0$ in Definition 7.24. If no such modulus exists, we must choose another base vector.

(iii) $$M_{-r} = \prod_k m_k$$

where m_k is a modulus for which $r_k < 0$ in Definition 7.24. If no such modulus exists, $M_{-r} = 1$.

Obviously,

(7.33) $$M = M_r M_0 M_{-r}.$$

7.34. **Definition.** Let $|a/b|_\beta$ be given. Then q in $\{0, 1, \ldots, M_0 - 1\}$ is the integer which has the property that, for each modulus m_j in M_0,

$$|q|_{m_j}^* = \left|\frac{a}{b}\right|_{m_j}^*.$$

7.35. **Definition.**

$$q' = \left|qM_{-r}M_r^{-1}(M_0)\right|_{M_0}.$$

The following theorem gives us an algorithm for mapping $\tilde{\mathbb{T}}_\beta$ onto $\mathbb{F}_N$.

7.36. **Theorem.** *Given $|a/b|_\beta$ in $\tilde{\mathbb{T}}_\beta$, we can find a unique order-N Farey fraction a/b by using the inverse mapping of Section 6, with the seed matrix*

$$\begin{bmatrix} M_r M_0 & 0 \\ M_r q' & M_{-r} \end{bmatrix}.$$

PROOF. See the forthcoming paper by Matula and Gregory mentioned above. □

7.37. EXAMPLE. Let $\beta = [2, 3, 5, 7]$ which implies $M = 210$ and $N = 10$. We will show that

$$[(1, 0), (1, 1), (4, 0), (3, -1)]$$

maps onto $\frac{3}{7}$ in $\mathbb{F}_{10}$.

  (i) First, we compute $M_r = 3$, $M_0 = 10$, and $M_{-r} = 7$.
 (ii) Next, we find q, the integer in $\{0, 1, \ldots, 9\}$ for which

$$|q|_2^* = (1, 0),$$
$$|q|_5^* = (4, 0).$$

In other words, we find the integer q for which

$$|q|_2 = 1,$$
$$|q|_5 = 4.$$

Using the mixed-radix procedure exhibited in Problem 7.19, we obtain

$$q = 9.$$

(iii) $M_r^{-1}(M_0) = 3^{-1}(10) = 7.$
(iv) $q' = |9 \cdot 7 \cdot 7|_{10} = 1.$

(v) Using these values, we obtain the seed matrix

$$\begin{bmatrix} M_r M_0 & 0 \\ M_r q' & M_{-r} \end{bmatrix} = \begin{bmatrix} 30 & 0 \\ 3 & 7 \end{bmatrix}.$$

With this seed matrix, we obtain

$$
\begin{array}{c|cc}
 & 30 & 0 \\
 & 3 & 7 \\
\hline
10 & 0 & -70
\end{array}
$$

and we have shown that

$$[(1, 0), (1, 1), (4, 0), (3, -1)] \to \tfrac{3}{7}.$$

Arithmetic in $\tilde{\mathbb{T}}_\beta$

The numbers of the set $\tilde{\mathbb{T}}_\beta$ are n-tuples whose elements are ordered pairs (see Definition 7.24). Arithmetic operations on these n-tuples are performed in an elementwise manner according to the rules described below. However, before the rules are stated, we must introduce some notation.

For $i = 1, 2, \ldots, n$, we use (7.21) and (7.22) to write

(7.38)
$$\frac{a}{b} = \frac{a_i}{b_i} \cdot (m_i)^{r_i}$$

and

(7.39)
$$\frac{c}{d} = \frac{c_i}{d_i} \cdot (m_i)^{s_i}.$$

Also, for $i = 1, 2, \ldots, n$, let

(7.40)
$$f_i = \left| \frac{a_i}{b_i} \right|_{m_i}$$

and

(7.41)
$$g_i = \left| \frac{c_i}{d_i} \right|_{m_i}.$$

Then the (elementwise) rule for multiplication is

(7.42)
$$(f_i, r_i)(g_i, s_i) = (|f_i g_i|_{m_i}, r_i + s_i),$$

for $i = 1, 2, \ldots, n$.

7.43. EXAMPLE. We show that $\tfrac{3}{7} \cdot \tfrac{7}{3} = 1$, using the results in Examples (7.25) and 7.27, where $\beta = [2, 3, 5, 7]$.

$$[(1, 0), (1, 1), (4, 0), (3, -1)]$$
$$\boxdot [(1, 0), (1, -1), (4, 0), (5, 1)]$$
$$\overline{[(1, 0), (1, 0), (1, 0), (1, 0)]}$$

This is the representation for 1.

The (elementwise) rule for addition is a bit more complicated than the rule for multiplication. To express $(f_i, r_i) + (g_i, s_i)$ we introduce the notation

(7.44) $$h_i = |f_i + g_i|_{m_i}$$

for $i = 1, 2, \ldots, n$. The rule for addition is best described by using the following table. Notice that there are only four types of elements (ordered pairs) which must be considered; $(0, z)$ and those elements for which the first component is nonzero and the second component is positive, negative, or zero. We assume that r_i and s_i are positive and $f_i g_i \neq 0$.

7.45. Table. The Addition Table for Ordered Pairs

+	$(0, z)$	(f_i, r_i)	$(f_i, 0)$	$(f_i, -r_i)$
$(0, z)$	$(0, z)$	(f_i, r_i)	$(f_i, 0)$	$(f_i, -r_i)$
(g_i, s_i)	(g_i, s_i)	$\begin{cases} (h_i, r_i), \ r_i = s_i \\ (g_i, s_i), \ r_i > s_i \\ (f_i, r_i), \ r_i < s_i \end{cases}$	$(f_i, 0)$	$(f_i, -r_i)$
$(g_i, 0)$	$(g_i, 0)$	$(g_i, 0)$	$(h_i, 0)$	$(f_i, -r_i)$
$(g_i, -s_i)$	$(g_i, -s_i)$	$(g_i, -s_i)$	$(g_i, -s_i)$	$\begin{cases} (h_i, -r_i), \ r_i = s_i \\ (f_i, -r_i), \ r_i > s_i \\ (g_i, -s_i), \ r_i < s_i \end{cases}$

7.46. EXAMPLE. We show that $\frac{3}{7} - \frac{3}{7} = 0$, using the results in Examples 7.25 and 7.29, where $\beta = [2, 3, 5, 7]$.

$$[(1, 0), (1, 1), (4, 0), (3, -1)]$$
$$\boxplus [(1, 0), (2, 1), (1, 0), (4, -1)]$$
$$\overline{[(0, 0), (0, 1), (0, 0), (0, -1)]}$$

Since the first component in each ordered pair is 0, we know (from Definition 7.24) that this is a representation for zero.

7.47. EXAMPLE. We show that $\frac{3}{7} + 1 = \frac{10}{7}$, using $\beta = [2, 3, 5, 7]$.

$$[(1, 0), (1, 1), (4, 0), (3, -1)]$$
$$\boxplus [(1, 0), (1, 0), (1, 0), (1, 0)]$$
$$\overline{[(0, 0), (1, 0), (0, 0), (3, -1)]}$$

(i) First, we compute $M_r = 1$, $M_0 = 30$, and $M_{-r} = 7$.
(ii) Next, we find $q = 10$.
(iii) Since $M_r = 1$, we have $M_r^{-1} = 1$.
(iv) Then, we compute $q' = |10 \cdot 7 \cdot 1|_{30} = 10$.

(v) Finally,

$$\begin{bmatrix} M_r M_0 & 0 \\ M_r q' & M_{-r} \end{bmatrix} = \begin{bmatrix} 30 & 0 \\ 10 & 7 \end{bmatrix}.$$

Hence, with this seed matrix,

$$
\begin{array}{c|cc}
 & 30 & 0 \\
\hline
 & 10 & 7 \\
\hline
3 & 0 & -21
\end{array}
$$

and we have shown that

$$[(0, 0), (1, 0), (0, 0), (3, -1)] \to \tfrac{10}{7}.$$

7.48. Remark. If we use Definition 7.24, then

$$\tfrac{10}{7} \to [(1, 1), (1, 0), (1, 1), (3, -1)].$$

This result might be termed the *canonical representation* for $\tfrac{10}{7}$ and the result obtained in Example 7.47 (as a result of addition) might be termed a *non-canonical representation* for $\tfrac{10}{7}$. Notice that both representations are correctly mapped onto $\tfrac{10}{7}$.

The relationship between the canonical representation and the non-canonical representations for a rational number will be discussed in detail in the forthcoming paper by Matula and Gregory.

7.49. Remark. Obviously, a practical implementation of multiple-modulus residue arithmetic would involve the use of many very large prime moduli (as large as the length of a word in a fixed-word-length computer). The scheme is ideal for a parallel computer, because the elementwise operations can be carried out simultaneously using one processor for each modulus.

EXERCISES I.7

1. Evaluate x, using multiple-modulus residue arithmetic with $\beta = [11, 13]$, if

$$x = \tfrac{1}{2} - \tfrac{2}{3} - \tfrac{1}{6}.$$

2. Evaluate the expression for x in Problem 7, Exercises I.6, using multiple-modulus residue arithmetic with $\beta = [23, 29]$.

3. Prove Theorem 7.8.

4. Prove Corollary 7.9.

5. Prove Theorem 7.11.

6. Repeat Problem 1 with $\beta = [2, 3, 5, 7]$.

7. Repeat Problem 2 with $\beta = [3, 5, 7, 11]$.

CHAPTER II
Finite-Segment p-adic Arithmetic

1 Introduction

In this chapter we are motivated to present an alternative number system, the finite-segment p-adic number system introduced by Krishnamurthy, Rao, and Subramanian [1975a], [1975b], and by Alparslan [1975]. This number system is finite and its relation to the (infinite) p-adic number system of Hensel [1908] is explained in subsequent sections.

From a mathematical point of view, finite-segment p-adic arithmetic is equivalent to single-modulus residue arithmetic if the single modulus is an integer of the form $m = p^r$, where p is a prime and r is a positive integer. However, there are advantages in the p-adic representation when we desire to extend the discussion from rational numbers to polynomials.

In this number system, each rational number in a finite set (the order-N Farey fractions introduced in Chapter I) is mapped onto a unique coded representation called its Hensel code,* and arithmetic operations on pairs of rational numbers in this set can be replaced by corresponding arithmetic operations on their Hensel codes. It should be pointed out that finite-segment p-adic arithmetic shares with residue arithmetic the fact that it is free of rounding errors.

2 The Field of p-adic Numbers

Let $\mathbb{K}$ be an arbitrary field and let $\mathbb{R}$ be the field of real numbers. We define a *norm* (sometimes called a *valuation*) on $\mathbb{K}$ by the following mapping.

---

*Named for the German mathematician K. Hensel (1861–1941).

2.1. **Definition**. A norm on a field $\mathbb{K}$ (considered a vector space over itself) is a mapping $\|\cdot\|: \mathbb{K} \to \mathbb{R}$ such that, for all α, β in $\mathbb{K}$,

(i) $\|\alpha\| \geq 0$,   and $\|\alpha\| = 0$ if and only if $\alpha = 0$,

(ii) $\|\alpha\beta\| = \|\alpha\| \cdot \|\beta\|$,

(iii) $\|\alpha + \beta\| \leq \|\alpha\| + \|\beta\|$.

For example, in the field of rational numbers $\mathbb{Q}$ the absolute value mapping $|\cdot|$ can be shown to be a norm on $\mathbb{Q}$. Another norm on $\mathbb{Q}$ (of more interest to us here) can be constructed on the observation that if $\alpha = a/b$ is a non-zero element of $\mathbb{Q}$ with $\gcd(a, b) = 1$, then α can be expressed uniquely in the form

$$(2.2) \qquad\qquad \alpha = \frac{c}{d} \cdot p^e,$$

where p is a given prime, where c, d, and e are integers with $\gcd(c, d) = 1$, and where p divides neither c nor d. With α defined in (2.2) we have the following result.

2.3. **Theorem**. *The mapping* $\|\cdot\|_p : \mathbb{Q} \to \mathbb{R}$ *defined by*

$$\|\alpha\|_p = \begin{cases} p^{-e} & \text{if } \alpha \neq 0 \\ 0 & \text{if } \alpha = 0 \end{cases}$$

is a norm on $\mathbb{Q}$.

PROOF. See Koblitz [1977], p. 2.   □

2.4. **Definition**. The norm in Theorem 2.3 is called the p-adic norm on $\mathbb{Q}$.

It should be observed that the p-adic norm is counter-intuitive since a large positive integer e in (2.2) implies a small value for the p-adic norm.

A metric space

Before continuing we need to introduce the concept of a *metric* and the concept of a *metric space*.

2.5. **Definition**. A metric space is a pair $(\mathbb{H}, d)$ consisting of a nonempty set $\mathbb{H}$ and a metric (or distance function) $d : \mathbb{H} \times \mathbb{H} \to \mathbb{R}$ such that, for all x, y, z in $\mathbb{H}$,

(i) $d(x, y) = 0$,   if and only if $x = y$,

(ii) $d(x, y) = d(y, x)$,

(iii) $d(x, z) \leq d(x, y) + d(y, z)$.

The properties (i), (ii), and (iii) are sometimes called the Hausdorff postulates* and it is not difficult to deduce from them a fourth property that, for all x, y in $\mathbb{H}$,

(iv) $$d(x, y) \geq 0.$$

2.6. **Definition.** A sequence $\{x_n\} = \{x_1, x_2, \ldots\}$, where $x_n \in \mathbb{H}$ for all n, is called a Cauchy sequence† in the metric space $(\mathbb{H}, d)$ if and only if

$$d(x_n, x_m) \to 0 \qquad (m, n \to \infty),$$

that is, for every $\varepsilon > 0$ there exists $N = N(\varepsilon)$ such that for all $n, m > N$,

$$d(x_n, x_m) < \varepsilon.$$

2.7. **Definition.** A sequence $\{x_n\} = \{x_1, x_2, \ldots\}$ in the metric space $(\mathbb{H}, d)$ is called *convergent* (to x) if and only if there exists $x \in \mathbb{H}$ such that

$$d(x_n, x) \to 0 \qquad (n \to \infty).$$

We then write $x_n \to x$ and call x the *limit* of the sequence.

Notice that there is nothing in the definition of a Cauchy sequence in a metric space which implies convergence.‡ In fact, it is well known that not every Cauchy sequence in a metric space is convergent. However, if a metric space has the property that every Cauchy sequence in the space converges, the metric space is given a special designation.

2.8. **Definition.** A metric space $(\mathbb{H}, d)$ is called *complete* if and only if every Cauchy sequence converges (to a point in $\mathbb{H}$). To be more specific, we require that if

$$d(x_n, x_m) \to 0 \qquad (n, m \to \infty),$$

then there exists $x \in \mathbb{H}$ such that

$$d(x_n, x) \to 0 \qquad (n \to \infty).$$

A particular metric space

We can construct a metric space by letting $\mathbb{H} = \mathbb{Q}$ and by defining a metric $d : \mathbb{Q} \times \mathbb{Q} \to \mathbb{R}$ in terms of the p-adic norm on $\mathbb{Q}$. Thus, if we define

(2.9) $$d(\alpha, \beta) = \|\alpha - \beta\|_p$$

for all α, β in $\mathbb{Q}$, then $(\mathbb{Q}, d)$ constitutes a metric space.

---

*Named for the German mathematician F. Hausdorff (1868–1942).

†Named for the French mathematician A. L. Cauchy (1789–1857).

‡Convergence means convergence to a point in the space.

2.10. Definition. The metric (2.9), induced by the p-adic norm $\|\cdot\|_p$, is called the *p-adic metric*.

Of particular interest in the metric space $(\mathbb{Q}, d)$ is the sequence of powers of the prime p

(2.11) $$\{p^n\} = \{p, p^2, p^3, \ldots\}.$$

It is interesting to note that this sequence converges to 0 because, in terms of the p-adic metric,

(2.12)
$$\begin{aligned} d(p^n, 0) &= \|p^n\|_p \\ &= p^{-n}, \end{aligned}$$

and $p^{-n} \to 0$ as $n \to \infty$. (Recall that the p-adic norm is counter-intuitive.)

Completion of a metric space

In the theory of metric spaces it is well known that for a non-complete metric space (not every Cauchy sequence converges) it is possible to construct a complete metric space (every Cauchy sequence converges) called the completion of the metric space. See Koblitz [1977], for example.

2.13. EXAMPLE. Consider the metric space $(\mathbb{Q}, \hat{d})$, where $\hat{d}$ is the absolute value metric

$$\hat{d}(\alpha, \beta) = |\alpha - \beta|.$$

Let $\hat{\mathbb{S}}$ be the set of Cauchy sequences in this metric space. We define two Cauchy sequences $s_1 = \{a_1, a_2, \ldots\}$ and $s_2 = \{b_1, b_2, \ldots\}$ to be *equivalent* (and we write $s_1 \sim s_2$) if and only if $|a_i - b_i| \to 0$ as $i \to \infty$. This is an *equivalence relation*, that is, $\sim$ has the following properties:

(i) $s_1 \sim s_1,$

(ii) $s_1 \sim s_2$ implies $s_2 \sim s_1,$

(iii) $s_1 \sim s_2$ and $s_2 \sim s_3$ imply $s_1 \sim s_3.$

We say that two sequences s_1 and s_2 belong to the same *equivalence class* if $s_1 \sim s_2$. We now define $\mathbb{R}$ to be the set of equivalence classes of Cauchy sequences in $(\mathbb{Q}, \hat{d})$. It is possible to define addition, multiplication and finding additive and multiplicative inverses of equivalence classes of Cauchy sequences in such a way that $(\mathbb{R}, +, \cdot)$ is a field. (See Koblitz [1977], for example.) This field is the *field of real numbers* and the metric space $(\mathbb{R}, \hat{d})$ is the completion of the metric space $(\mathbb{Q}, \hat{d})$.

We obtain the *field of p-adic numbers* if we find the completion of the rational numbers with respect to the p-adic metric rather than the absolute value metric used above. In this case we begin with the metric space $(\mathbb{Q}, d)$, where d is defined in (2.9). We let $\mathbb{Q}_p$ denote the set of equivalence classes of Cauchy sequences in $(\mathbb{Q}, d)$ where s_1 and s_2 are equivalent if and only if $\|a_i - b_i\|_p \to 0$ as $i \to \infty$. With addition, multiplication and finding inverses properly defined (see Koblitz [1977], p. 10), the system $(\mathbb{Q}_p, +, \cdot)$ constitutes a field, the field of p-adic numbers, and the metric space $(\mathbb{Q}_p, d)$ is the completion of the metric space $(\mathbb{Q}, d)$.

2.14. Definition. Each element of the set $\mathbb{Q}_p$ is called a p-adic number.

We have introduced the set of p-adic numbers $\mathbb{Q}_p$ in a rather abstract way. Perhaps the following expansion theorem for p-adic numbers will characterize them more concretely. This expansion is somewhat analogous to the decimal expansion for the real numbers.

2.15. Theorem. *Any p-adic number $\alpha \in \mathbb{Q}_p$ can be written in the form*

$$\alpha = \sum_{j=n}^{\infty} a_j p^j,$$

where each $a_j \in \mathbb{I}$, and n is such that $\|\alpha\|_p = p^{-n}$. Moreover, if we choose each a_j in $\{0, 1, 2, \ldots, p - 1\}$, then the expansion is unique. (In this case, the expansion is the canonical representation of α.)

PROOF. See Bachman [1964], pp. 34–35. □

Since the field of rational numbers is imbedded in the field of p-adic numbers, and since we are interested primarily in the p-adic expansion of a rational number, the following corollary is of interest.

2.16. Corollary. *Any rational number $\alpha \in \mathbb{Q}$ has the unique p-adic expansion*

$$\alpha = \sum_{j=n}^{\infty} a_j p^j$$

where each coefficient a_j is an integer in $\{0, 1, 2, \ldots, p - 1\}$ and n is such that $\|\alpha\|_p = p^{-n}$. (The infinite series converges to the rational number α in the p-adic metric.)

PROOF. The corollary is a direct consequence of Theorem 2.15. □

2.17. EXAMPLE. Consider the following power series expansion.

$$\alpha = 2 + 3p + p^2 + 3p^3 + p^4 + 3p^5 + \cdots$$
$$= 2 + 3p(1 + p^2 + p^4 + \cdots) + p^2(1 + p^2 + p^4 + \cdots)$$
$$= 2 + (3p + p^2)(1 + p^2 + p^4 + \cdots).$$

Since $1 + p^2 + p^4 + \cdots$ converges to $(1 - p^2)^{-1}$ in the p-adic metric (see Problem 5) we can write

$$\alpha = 2 + \frac{3p + p^2}{1 - p^2},$$

and if $p = 5$, we obtain

$$\alpha = \tfrac{1}{3}.$$

This is usually written in the abbreviated form

$$\tfrac{1}{3} = .23131313 \cdots \qquad (p = 5),$$

where the "point" at the left is called the p-adic point.

2.18. Remark. Notice that there is a one-to-one correspondence between the power series expansion

$$\alpha = a_n p^n + a_{n+1} p^{n+1} + a_{n+2} p^{n+2} + \cdots$$

and the abbreviated representation

$$\alpha = a_n a_{n+1} a_{n+2} \cdots$$

where only the coefficients of the powers of p are exhibited. Because of this correspondence we can use the power series expansion and the abbreviated representation interchangeably. In fact, we shall refer to each of them as the p-adic expansion for α. The abbreviated representation is completely analogous to the representation of the decimal expansion of a real number. In fact, we complete the analogy by introducing a p-adic point as a device for displaying the sign of n. Thus, we write

$$\alpha = \begin{cases} a_n a_{n+1} \cdots a_{-2} a_{-1} . a_0 a_1 a_2 \cdots & \text{for } n < 0 \\ .a_0 a_1 a_2 \cdots & \text{for } n = 0 \\ .0 \cdots 0 a_n a_{n+1} \cdots & \text{for } n > 0. \end{cases}$$

We know that a real number is rational if and only if its decimal expansion is periodic. Similarly, a p-adic number is rational if and only if its p-adic expansion is periodic. (See Bachman [1964], p. 40.) Consequently, since we are primarily interested in the p-adic expansions of rational numbers we will be dealing only with p-adic expansions which are periodic.

The complement representation for a negative rational number

The relationship between the p-adic expansions for α and for $-\alpha$ can be seen in the following result.

2.19. Theorem. *If*

$$\alpha = a_n p^n + a_{n+1} p^{n+1} + a_{n+2} p^{n+2} + \cdots,$$

then

$$-\alpha = b_n p^n + b_{n+1} p^{n+1} + b_{n+2} p^{n+2} + \cdots,$$

where $b_n = p - a_n$ and $b_j = (p-1) - a_j$ for $j > n$.

PROOF. We sketch a proof by showing that the sum of the two representations is zero. Write

$$-\alpha = (p - a_n)p^n + (p - 1 - a_{n+1})p^{n+1} + (p - 1 - a_{n+2})p^{n+2} + \cdots$$

If we form $\alpha + (-\alpha)$ we obtain

$$0 = p : p^n + (p - 1) \cdot p^{n+1} + (p - 1) \cdot p^{n+2} + \cdots$$
$$= \quad 0 + \qquad p \cdot p^{n+1} + (p - 1) \cdot p^{n+2} + \cdots$$
$$= \quad 0 + \qquad 0 \quad + \qquad p \cdot p^{n+2} + \cdots$$
$$= \quad 0 + \qquad 0 \quad + \qquad 0 \quad + \cdots,$$

with zeros as far as we wish to carry the expansion.   □

2.20. EXAMPLE. Recall the 5-adic expansion for 1/3 in Example 2.17.

$$\tfrac{1}{3} = 2 + 3 \cdot 5 + 1 \cdot 5^2 + 3 \cdot 5^3 + 1 \cdot 5^4 + \cdots.$$

Hence,

$$-\tfrac{1}{3} = 3 + 1 \cdot 5 + 3 \cdot 5^2 + 1 \cdot 5^3 + 3 \cdot 5^4 + \cdots$$

and, when we add, we obtain

$$0 = 5 + 4 \cdot 5 + 4 \cdot 5^2 + 4 \cdot 5^3 + \cdots$$
$$= 0 + 5 \cdot 5 + 4 \cdot 5^2 + 4 \cdot 5^3 + \cdots$$
$$= 0 + \quad 0 + 5 \cdot 5^2 + 4 \cdot 5^3 + \cdots$$
$$= 0 + \quad 0 + \quad 0 + 5 \cdot 5^3 + \cdots$$
$$= 0 + \quad 0 + \quad 0 + \quad 0 + \cdots.$$

The p-adic representation of a radix fraction

If $\alpha = a/b$ is a rational number with $(a, b) = 1$ and if b is a power of p, we call α a *radix fraction*. It is interesting to observe what happens when the radix fraction is positive.

2.21. **Theorem.** *If $\alpha \in \mathbb{Q}_p$, then α can be represented by a finite* p-adic expansion if and only if α is a positive radix fraction.*

PROOF. See MacDuffee [1938].   □

*A p-adic expansion is finite if it contains only a finite number of nonzero p-adic digits.

2.22. Corollary. *If α is a positive integer, then α can be represented by a finite p-adic expansion.*

PROOF. If α is a positive integer, it is automatically a positive radix fraction.

$\square$

Notice that when we complement a finite p-adic expansion, using Theorem 2.19, we automatically get an infinite p-adic expansion. Hence, the theorem and corollary above refer only to positive radix fractions (and integers). For example,

$$(2.23) \qquad \frac{199}{125} = 442.100000\ldots \qquad (p = 5),$$

and

$$(2.24) \qquad -\frac{199}{125} = 102.344444\ldots \qquad (p = 5).$$

It is easily verified that

$$(2.25) \qquad 199 = .442100000\ldots \qquad (p = 5)$$

and

$$(2.26) \qquad -199 = .10234444\ldots \qquad (p = 5)$$

which illustrates the fact that if $\alpha = a/b$, with $\gcd(a, b) = 1$ and $b = p^k$, then the p-adic representation of α can be obtained from the p-adic representation of a merely by shifting the p-adic point k places to the right. This is completely analogous to the situation that exists with decimal fractions whose denominators are powers of ten.

2.27. Remark. Since a positive integer h can be expressed in exactly one way as the sum of powers of a prime p if the coefficients are integers in the set $\mathbb{I}_p$, that is, since

$$h = d_0 + d_1 p^2 + \cdots + d_k p^k$$

with the integers d_i in the set $\mathbb{I}_p$, we see that there is essentially no difference between the radix-p (or p-ary) representation of h and the p-adic representation of h. In fact, the only difference is that, in the abbreviated representations, *the digits are written in reverse order*. For example

$$14 = 2 + 3 + 3^2$$

which means that the radix-3 representation is

$$14_{\text{ten}} = 112_{\text{three}}.$$

However, the 3-adic representation is

$$14_{\text{ten}} = .2110000\ldots \qquad (p = 3)$$

and (since the number of non-zero digits is finite) we usually write

$$14_{\text{ten}} = .211 \qquad (p = 3).$$

Notice that the radix-3 representation and the 3-adic representation are mirror images of each other.

Computing the digits in a *p*-adic expansion

Let α have the *p*-adic expansion

$$\alpha = a_n p^n + a_{n+1} p^{n+1} + a_{n+2} p^{n+2} + \cdots$$

(2.28)
$$= p^n(a_n + a_{n+1} p + a_{n+2} p^2 + \cdots)$$

$$= p^n\left(\frac{c_1}{d_1}\right)$$

where $\gcd(c_1, d_1) = 1$, and p divides neither c_1 nor d_1. The *p*-adic expansion for c_1/d_1 is

(2.29)
$$\frac{c_1}{d_1} = a_n + a_{n+1} p + a_{n+2} p^2 + \cdots$$

and so

(2.30)
$$\left|\frac{c_1}{d_1}\right|_p = |a_n + a_{n+1} p + a_{n+2} p^2 + \cdots|_p$$

$$= a_n.$$

In other words, we obtain a_n by computing

(2.31)
$$a_n = \left|\frac{c_1}{d_1}\right|_p.$$

Next, we use (2.29) to form the expression

$$\frac{c_1}{d_1} - a_n = p(a_{n+1} + a_{n+2} p + a_{n+3} p^2 + \cdots)$$

(2.32)
$$= p\left(\frac{c_2}{d_2}\right)$$

where $\gcd(c_2, d_2) = 1$. The *p*-adic expansion for c_2/d_2 is

(2.33)
$$\frac{c_2}{d_2} = a_{n+1} + a_{n+2} p + a_{n+3} p^2 + \cdots$$

and so

(2.34)
$$\left|\frac{c_2}{d_2}\right|_p = |a_{n+1} + a_{n+2} p + a_{n+3} p^2 + \cdots|_p$$

$$= a_{n+1}.$$

In other words, we obtain a_{n+1} by computing

$$(2.35) \qquad\qquad a_{n+1} = \left| \frac{c_2}{d_2} \right|_p .$$

In general, we continue this procedure and, unless α is a positive radix fraction (see Theorem 2.21), the process will not terminate. However, since we are assuming that α is a rational number, the p-adic expansion will be periodic and we need continue only until the period has been exhibited.

2.36. EXAMPLE. Let $\alpha = 2/3$ and let $p = 5$. In this case $c_1 = 2$, $d_1 = 3$ and $n = 0$. Thus, using the forward mapping of Chapter I, we obtain

$$a_0 = \left| \tfrac{2}{3} \right|_5$$

$$= 4.$$

Next we form the expression

$$\frac{c_1}{d_1} - a_0 = \frac{2}{3} - 4$$

$$= 5 \left(\frac{-2}{3} \right).$$

In this case $c_2 = -2$ and $d_2 = 3$. Thus,

$$a_1 = \left| \tfrac{-2}{3} \right|_5$$

$$= 1.$$

Now we form the expression

$$\frac{c_2}{d_2} - a_1 = \frac{-2}{3} - 1$$

$$= 5 \left(\frac{-1}{3} \right).$$

In this case $c_3 = -1$ and $d_3 = 3$. Thus,

$$a_2 = \left| \tfrac{-1}{3} \right|_5$$

$$= 3.$$

If we continue this procedure, we obtain $a_3 = 1$ and $a_4 = 3$. At this point we may stop because the period of the p-adic expansion has been revealed. Hence,

$$\tfrac{2}{3} = .4131313\ldots \qquad (p = 5).$$

2.37. **Remark.** We have pointed out some of the similarities between p-adic numbers and decimal numbers. However, one of the differences between p-adic numbers and decimal numbers is that if two p-adic expansions converge to the same number in $\mathbb{Q}_p$, they are the same number, that is, *all* *of their digits are the same.* Thus, we never encounter a situation analogous to the example $1 = 0.9999\ldots$, in which a terminating decimal can also be represented by a non-terminating decimal with repeating nines.

EXERCISES II.2

1. Consider the field of rational numbers $(\mathbb{Q}, +, \cdot)$. Show that the absolute-value mapping $|\cdot| : \mathbb{Q} \to \mathbb{R}$ is a norm on $\mathbb{Q}$.

2. (a) What is the radix-5 representation of 14?
   (b) What is the 5-adic representation of 14?
   (c) What is the 5-adic representation of -14?

3. Compute the digits in the 5-adic expansion of
   (a) $\alpha = \frac{4}{3}$ (d) $\alpha = \frac{1}{6}$
   (b) $\alpha = \frac{5}{2}$ (e) $\alpha = -9$
   (c) $\alpha = -\frac{5}{2}$ (f) $\alpha = -10$

4. Use Definition 2.5 to prove that $d(x, y) \geq 0$.

5. Show that $1 + p^2 + p^4 + \cdots$ converges to $(1 - p^2)^{-1}$ in the p-adic metric.

3 Arithmetic in $\mathbb{Q}_p$

The operations of addition, subtraction, multiplication and division in $\mathbb{Q}_p$ are quite similar to the corresponding operations on decimals. The main difference, however, is that we proceed from "left to right" rather than from "right to left" as we do with decimals.

Addition

In Example 2.20 we demonstrated how to add $\frac{1}{3}$ to $-\frac{1}{3}$ using their power series expansions. In general, suppose we have two arbitrary p-adic numbers

$$(3.1) \qquad \alpha = a_n p^n + a_{n+1} p^{n+1} + a_{n+2} p^{n+2} + \cdots$$

and

$$(3.2) \qquad \beta = b_n p^n + b_{n+1} p^{n+1} + b_{n+2} p^{n+2} + \cdots,$$

where the p-adic digits a_i and b_i lie in the set $\mathbb{I}_p$. We point out that either a_n or b_n may be zero but not both. There are no conditions imposed on subsequent digits.

If we add α and β we obtain

$$(3.3) \qquad \begin{aligned} \alpha + \beta &= (a_n + b_n)p^n + (a_{n+1} + b_{n+1})p^{n+1} + (a_{n+2} + b_{n+2})p^{n+2} + \cdots \\ &= c_n p^n + c_{n+1} p^{n+1} + c_{n+2} p^{n+2} + \cdots \end{aligned}$$

where

(3.4) $$c_i = a_i + b_i, \qquad i = n, n+1, \ldots .$$

Suppose $c_n, c_{n+1}, \ldots, c_{k-1}$ are digits (that is, suppose they are less than p) but c_k is not. Then

(3.5) $$c_k = p + d_k$$

where $0 \le d_k < p$. In this case,

$$\begin{aligned}\alpha + \beta = c_n p^n + &\cdots + c_{k-1}p^{k-1} + (p + d_k)p^k \\ &+ c_{k+1}p^{k+1} + \cdots \\ = c_n p^n + &\cdots + c_{k-1}p^{k-1} + d_k p^k \\ &+ (c_{k+1} + 1)p^{k+1} + \cdots\end{aligned}$$

(3.6)

and d_k is the digit associated with p^k. Notice that a "carry" has been generated and so c_{k+1} is increased by one unit. At this point $c_{k+1} + 1$ must be examined to see whether or not it is less than p. If it is not, we have a "carry" propagated to c_{k+2} and so on.

This situation is similar to the situation which exists when we add decimals. However, in the p-adic case we add the digits (using radix-p arithmetic) and, from the point of view of the abbreviated representation, we proceed from left to right (rather than from right to left as in the case of decimals).

3.7. EXAMPLE. Add $\frac{2}{3}$ and $\frac{5}{6}$ in $\mathbb{Q}_5$. It is easy to verify that the p-adic expansions for the two operands are

$$\tfrac{2}{3} = .4131313\ldots \qquad (p = 5)$$

and

$$\tfrac{5}{6} = .0140404\ldots \qquad (p = 5).$$

Now if we add the p-adic expansions, using radix-5 arithmetic (proceeding from left to right), we obtain

$$\begin{array}{r} .4131313\ldots \\ .0140404\ldots \\ \hline .4222222\ldots \end{array}$$

As a check we observe that $\frac{2}{3} + \frac{5}{6} = \frac{3}{2}$ and

$$\tfrac{3}{2} = .4222222\ldots \qquad (p = 5).$$

Subtraction

In order to do subtraction we complement the subtrahend (using Theorem 2.19) and add it to the minuend, that is, $\alpha - \beta = \alpha + (-\beta)$.

3.8. **EXAMPLE.** Subtract $\frac{5}{6}$ from $\frac{2}{3}$ in $\mathbb{Q}_5$. If we complement the representation for $\frac{5}{6}$ in the previous example, we see that

$$-\tfrac{5}{6} = .0404040\ldots \qquad (p = 5).$$

Thus, if we add the p-adic expansions for $\frac{2}{3}$ and $-\frac{5}{6}$, using radix-5 arithmetic (proceeding from left to right), we obtain

$$
\begin{array}{l}
.4131313\ldots \\
.0404040\ldots \\
\hline
.4040404\ldots
\end{array}
$$

As a check we observe that $\frac{2}{3} - \frac{5}{6} = -\frac{1}{6}$ and

$$-\tfrac{1}{6} = .4040404\ldots \qquad (p = 5).$$

Multiplication

Before discussing multiplication we note that we can always write a p-adic number γ in the form

$$
\begin{aligned}
\gamma &= g_n p^n + g_{n+1} p^{n+1} + g_{n+2} p^{n+2} + \cdots \\
(3.9) \qquad &= p^n (g_n + g_{n+1} p + g_{n+2} p^2 + \cdots) \\
&= p^n \alpha
\end{aligned}
$$

where α has the obvious definition, and n can be positive, negative, or zero. If we change the notation so that the subscripts on the p-adic digits match the exponents on p, we can write

$$(3.10) \qquad \alpha = a_0 + a_1 p + a_2 p^2 + \cdots.$$

3.11. **Definition.** Any p-adic number whose p-adic expansion contains no negative powers of p is called a p-adic *integer*. Any p-adic integer whose first digit is non-zero is called a p-adic *unit*.

Thus, in (3.10), α is a p-adic unit and any p-adic number can be written as a product of a p-adic unit and a power of p.

If we have γ given by (3.9) and

$$(3.12) \qquad \delta = p^m \beta$$

where α and β are p-adic units, then

$$(3.13) \qquad \gamma\delta = p^{n+m} \alpha\beta.$$

Thus, there is no loss of generality in restricting our discussion of p-adic multiplication to a discussion of the multiplication of p-adic units.

Let α and β be the p-adic units

$$(3.14) \qquad
\begin{cases}
\alpha = a_0 + a_1 p + a_2 p^2 + \cdots \\
\beta = b_0 + b_1 p + b_2 p^2 + \cdots
\end{cases}
$$

(where $a_0 b_0 \neq 0$). Then

$$(3.15) \quad \alpha\beta = (a_0 + a_1 p + a_2 p^2 + \cdots)(b_0 + b_1 p + b_2 p^2 + \cdots)$$
$$= c_0 + c_1 p + c_2 p^2 + c_3 p^3 + \cdots$$

with

$$(3.16) \quad \begin{cases} c_0 = a_0 b_0 \\ c_1 = a_0 b_1 + a_1 b_0 \\ c_2 = a_0 b_2 + a_1 b_1 + a_2 b_0 \\ \vdots \\ c_k = a_0 b_k + a_1 b_{k-1} + \cdots + a_k b_0. \\ \vdots \end{cases}$$

For a matrix representation, see (3.23).

Even though the p-adic digits a_i and b_i lie in the set $\mathbb{I}_p$ we cannot assume that the integers c_i lie in this set, that is, we cannot assume that they are digits. (In general, they are not.) Hence, we write

$$(3.17) \quad \begin{aligned} c_0 &= a_0 b_0 \\ &= d_0 + t_1 p \end{aligned}$$

where $0 \leq d_0 < p$. Then d_0 is the first digit in the p-adic expansion for $\alpha\beta$ and t_1 is the "carry" which we must add to c_1. Next we write

$$(3.18) \quad \begin{aligned} c_1 + t_1 &= (a_0 b_1 + a_1 b_0) + t_1 \\ &= d_1 + t_2 p \end{aligned}$$

where $0 \leq d_1 < p$. Then d_1 is the second digit in the p-adic expansion for $\alpha\beta$ and t_2 is the "carry" which we must add to c_2. If we continue this procedure, we obtain the (unique) p-adic expansion

$$(3.19) \quad \alpha\beta = d_0 + d_1 p + d_2 p^2 + \cdots$$

where $0 \leq d_i < p$ for all i.

Again we point out that this situation is similar to the situation which exists when we multiply decimals. However, in the p-adic case the arithmetic indicated in (3.16) and in adding the "carry" in (3.18) is radix-p arithmetic and, from the point of view of the abbreviated representation, we proceed from left to right (rather than from right to left as in the case of decimals).

3.20. EXAMPLE. Multiply $\frac{2}{3}$ by $\frac{1}{6}$ in $\mathbb{Q}_5$. In Example 3.7 we find the 5-adic expansions for $\frac{2}{3}$ and $\frac{5}{6}$. It is easily verified that the 5-adic expansion for $\frac{1}{6}$ is obtained from the 5-adic expansion for $\frac{5}{6}$ by a single shift of the p-adic point. Hence,

$$\tfrac{1}{6} = .14040404\ldots \qquad (p = 5).$$

If we multiply these expansions, we obtain

$$
\begin{array}{r}
.4131313131313\ldots \\
.1404040404040\ldots \\
\hline
4131313131313\ldots \\
123131313131\ldots \\
1231313131\ldots \\
12313131\ldots \\
123131\ldots \\
1231\ldots \\
12\ldots \\
\hline
.4201243201243\ldots
\end{array}
$$

As a check we observe that $(\frac{2}{3})(\frac{1}{6}) = \frac{1}{9}$, and

$$\tfrac{1}{9} = .4\,201243\,201243\ldots \qquad (p = 5).$$

Division

Using an argument similar to that used for multiplication, there is no loss in generality in restricting our discussion of p-adic division to the division of p-adic units. Consequently, consider the p-adic units

$$
(3.21) \qquad
\begin{cases}
\delta = d_0 + d_1 p + d_2 p^2 + \cdots \\
\beta = b_0 + b_1 p + b_2 p^2 + \cdots
\end{cases}
$$

with $d_0 b_0 \neq 0$. The quotient $\alpha = \delta/\beta$ can be written

$$
(3.22) \qquad
\begin{aligned}
\alpha &= \frac{d_0 + d_1 p + d_2 p^2 + \cdots}{b_0 + b_1 p + b_2 p^2 + \cdots} \\
&= a_0 + a_1 p + a_2 p^2 + \cdots
\end{aligned}
$$

where we assume that $a_0, a_1, a_2, \ldots$ are digits.

To compute a_0 we observe that $\delta = \alpha\beta$ and if we form $\alpha\beta$, as in (3.15), we are led to the equations (3.16). These equations give us the coefficients $c_0, c_1, \ldots$ which are not digits and we must use (3.17) and (3.18) to get the digits $d_0, d_1, \ldots$ in (3.19). Observe that (3.16) has the equivalent matrix form

$$
(3.23) \qquad
\begin{bmatrix}
b_0 & & & & \\
b_1 & b_0 & & & \\
b_2 & b_1 & b_0 & & \\
\cdots & \cdots & \cdots & \cdots & \cdots \\
b_k & b_{k-1} & b_{k-2} & \cdots & b_0 \\
\cdots & \cdots & \cdots & \cdots & \cdots
\end{bmatrix}
\begin{bmatrix}
a_0 \\
a_1 \\
a_2 \\
\vdots \\
a_k \\
\vdots
\end{bmatrix}
=
\begin{bmatrix}
c_0 \\
c_1 \\
c_2 \\
\vdots \\
c_k \\
\vdots
\end{bmatrix}.
$$

In both (3.16) and (3.23) we have $a_0 b_0 = c_0$. However, from (3.17), we have $c_0 = d_0 + t_1 p$. Hence,

(3.24) $$a_0 b_0 = d_0 + t_1 p,$$

from which we obtain $|a_0 b_0|_p = d_0$. Therefore,

(3.25) $$a_0 = |d_0 b_0^{-1}|_p.$$

In other words, to get the first digit of the quotient we form $b_0^{-1}(p)$, multiply this by d_0, and reduce the result modulo p.

It turns out that this is the clue for obtaining each digit of the expansion for α in (3.22). At each stage of the standard "long division" procedure we multiply $b_0^{-1}(p)$ by the first digit of the partial remainder and reduce the result modulo p. (At the first step, the dividend is considered to be the partial remainder.)

3.26. EXAMPLE. Divide $\frac{2}{3}$ by $\frac{1}{12}$ in $\mathbb{Q}_5$. We have

$$\tfrac{2}{3} = .4131313\ldots \qquad (p = 5)$$

and

$$\tfrac{1}{12} = .3424242\ldots \qquad (p = 5).$$

In the "long division" which follows we do our subtraction at each step by complementing the subtrahend and adding.

The first digit of the divisor is $b_0 = 3$ and its multiplicative inverse is

$$b_0^{-1}(p) = 3^{-1}(5)$$
$$= 2.$$

The first digit of the partial remainder (in the first step this is the dividend) is $d_0 = 4$ which gives us

$$a_0 = |4 \cdot 2|_5$$
$$= 3.$$

Thus, the first step of the long division procedure is

$$
\begin{array}{r}
.3 \\
.3424242\ldots \overline{)\,.4131313\ldots} \\
1111111\ldots \\
\hline
342424\ldots
\end{array}
$$

and the new partial remainder is $342424\ldots$

To get the second digit in the quotient we multiply $b_0^{-1}(p)$ by the first digit of the partial remainder and reduce the result modulo p. Hence,

$$a_1 = |3 \cdot 2|_5$$
$$= 1.$$

Thus, the second step of the long division procedure gives us

$$
\begin{array}{r}
.31 \\
.3424242\ldots\overline{)\,.4131313\ldots} \\
1111111\ldots \\
\overline{342424\ldots} \\
202020\ldots \\
\overline{00000\ldots}
\end{array}
$$

In this particular example we have produced a partial remainder which is zero. Hence, we terminate the expansion at this point because all remaining digits of the quotient are zero. In general, this does not happen and we continue until the period in the expansion has been exhibited.

As a check we observe that $\frac{2}{3} \div \frac{1}{12} = 8$
and

$$8 = .3100000\ldots \qquad (p = 5).$$

3.27. **Remark.** We point out that division of *p*-adic numbers is deterministic and not subject to trial and error as is the case with the division of decimals. This is due to the fact that, once we have $b_0^{-1}(p)$, where b_0 is the first digit of the divisor, the algorithm for obtaining each digit of the quotient is very specific; we multiply the first digit of each partial remainder by $b_0^{-1}(p)$ and reduce the result modulo p.

EXERCISES II.3

In each problem do the arithmetic in $\mathbb{Q}_5$.

1. Add $\frac{4}{3}$ to $\frac{1}{6}$.

2. Add $\frac{1}{3}$ to $\frac{5}{2}$.

3. Subtract $\frac{4}{3}$ from $\frac{1}{6}$.

4. Subtract $\frac{1}{3}$ from $\frac{5}{2}$.

5. Multiply $\frac{4}{3}$ by $\frac{1}{6}$.

6. Multiply $\frac{1}{3}$ by $\frac{5}{2}$.

7. Divide $\frac{4}{3}$ by $\frac{1}{6}$.

8. Divide $\frac{1}{3}$ by $\frac{5}{2}$.

4 A Finite-Segment *p*-adic Number System

As we mentioned in the introduction to this chapter, a finite number system, based on the system of *p*-adic numbers described in the last two sections, has been proposed recently by Krishnamurthy, Rao, and Subramanian [1975a],

[1975b], and by Alparslan [1975]. In this number system, $\mathbb{F}_N$ (the set of order-N Farey fractions defined in Definition 5.13, Chapter I) is mapped onto the set of coded representations called Hensel codes, and arithmetic operations on the Hensel codes are equivalent to corresponding arithmetic operations on the order-N Farey fractions.

The integer N, which defines the set $\mathbb{F}_N$, is the largest positive integer which satisfies the inequality

(4.1) $$2N^2 + 1 \leq m$$

where

(4.2) $$m = p^r,$$

with p a prime, and r a positive integer.

Hensel codes

The Hensel code for a rational number is simply a finite segment of its (infinite) p-adic expansion, with the length of the segment determined by r. We use the notation $H(p, r, \alpha)$ to represent the Hensel code for $\alpha = a/b$, where r specifies the number of p-adic digits which we retain.

For example, since the (infinite) p-adic expansion for $\alpha = \frac{1}{3}$ is*

(4.3) $\frac{1}{3} = .2313131313\ldots$ $(p = 5)$,

the Hensel code for α, when $p = 5$ and $r = 4$, is

(4.4) $H(5, 4, \frac{1}{3}) = .2313.$

A table of Hensel codes $H(5, 4, \alpha)$ for the order-N Farey fractions α (satisfying $N = 17$) is given in Table 4.22.

The residue equivalent of a Hensel code

We now describe an algorithm for mapping a rational number α onto its Hensel code $H(p, r, \alpha)$ which does not involve finding its (infinite) p-adic expansion and truncating it to r digits. It is based on the following result.

4.5. Theorem. *Let* $\alpha = a/b$, *where*

$$\frac{a}{b} = \frac{c}{d} \cdot p^n$$

with $\gcd(c, d) = \gcd(c, p) = \gcd(d, p) = 1$. *Let the Hensel code for* c/d *be*

$$H(p, r, c/d) = .a_0 a_1 \ldots a_{r-1}.$$

*See Example 2.17 in this chapter.

Then $a_{r-1} \ldots a_1 a_0$ is the radix-p representation for the integer $\left| cd^{-1} \right|_m$ (with $m = p^r$). In other words,

$$\left| cd^{-1} \right|_m = a_0 + a_1 p + a_2 p^2 + \cdots + a_{r-1} p^{r-1}.$$

PROOF. Let c/d have the p-adic expansion

$$\frac{c}{d} = \sum_{j=0}^{\infty} a_j p^j$$

$$= (a_0 + a_1 p + \cdots + a_{r-1} p^{r-1}) + p^r R_r.$$

Then,

$$c = d(a_0 + a_1 p + \cdots + a_{r-1} p^{r-1}) + p^r(d \cdot R_r).$$

Hence, if $m = p^r$,

$$\left| c \right|_m = \left| d(a_0 + a_1 p + \cdots + a_{r-1} p^{r-1}) \right|_m$$

which implies

$$\left| cd^{-1} \right|_m = a_0 + a_1 p + \cdots + a_{r-1} p^{r-1}. \qquad \square$$

4.6. EXAMPLE. Let $p = 5$, $r = 4$, and $m = 625$. We compute the Hensel code for $\alpha = \frac{1}{3}$ as follows:

$$\left| \tfrac{1}{3} \right|_{625} = \left| 1 \cdot 3^{-1} \right|_{625}$$

$$= \left| 417 \right|_{625}$$

$$= 417$$

$$= 2 + 3 \cdot 5 + 1 \cdot 5^2 + 3 \cdot 5^3,$$

which implies

$$H(5, 4, \tfrac{1}{3}) = .2313.$$

In other words,

$$\left| \tfrac{1}{3} \right|_{625} = 417_{\text{ten}}$$

$$= 3132_{\text{five}}$$

and we obtain the digits of the Hensel code by writing the radix-5 digits in reverse order.

4.7. **Remark.** We can use Algorithm 6.26, Chapter I, in computing $\left| \tfrac{1}{3} \right|_{625}$. Hence, we construct the table

		625	0
		3	1
208		1	-208
	3	0	625

and obtain

$$3^{-1}(625) = |-208|_{625}$$

$$= 417.$$

4.8. Remark. The finite-segment p-adic representation (the Hensel code) for an order-N Farey fraction is equivalent to the single-modulus residue representation with the modulus $m = p^r$. In other words, what we are doing here is similar to what we did in Section 5, Chapter I. The primary differences are that we *always* choose $r > 1$ in this section, and after mapping an order-N Farey fraction onto an integer in $\mathbb{I}_m$, we then map the integer in $\mathbb{I}_m$ onto the corresponding Hensel code. The uniqueness of the Hensel code for each order-N Farey fraction follows from Theorem 5.14, Chapter I.

In Remark 2.18 we exhibited the p-adic point among the p-adic digits as a device for displaying the sign of n, when the rational number $\alpha = a/b$ is written in the form

(4.9)
$$\frac{a}{b} = \frac{c}{d} \cdot p^n,$$

with $\gcd(c, d) = \gcd(c, p) = \gcd(d, p) = 1$. When we use the first r digits in the p-adic representation in forming the Hensel code we keep the p-adic point in the same position. Consider the following three cases:

Case I

$$n = 0$$

In this case $a = c$ and $b = d$ in (4.9). The first step in finding $H(p, r, \alpha)$ is to compute the integer $|c \cdot d^{-1}|_m$. The second step is to express this (decimal) integer as a radix-p integer. The third step is to reverse the order of the digits.

For example, let $\alpha = 2/3$, $p = 5$, and $r = 4$. Then $c = 2$, $d = 3$, $n = 0$, and $p^r = 625$. Hence, if we use Algorithm 6.26, Chapter I, we obtain

(4.10)
$$\left|\tfrac{2}{3}\right|_{625} = 209.$$

Notice that 209 is the residue representation modulo 625. Next we write

(4.11)
$$209_{\text{ten}} = 1314_{\text{five}}.$$

If we write the radix-5 digits in reverse order, we obtain

(4.12)
$$H(5, 4, \tfrac{2}{3}) = .4131$$

and these four p-adic digits agree with the first four p-adic digits in Example 2.36.

Case II

$$n < 0$$

In this case $\alpha = (c/d)p^{-k}$, where k is a positive integer. To find $H(p, r, \alpha)$ we find $H(p, r, c/d)$, using the three steps in Case I, and then shift the *p*-adic point k places to the right.

For example, let $\alpha = \frac{2}{15}$, $p = 5$, and $r = 4$. We write $\alpha = (\frac{2}{3})5^{-1}$ which gives us $c = 2$, $d = 3$, $k = 1$, and $p^r = 625$. Since we already have $H(5, 4, \frac{2}{3})$ in (4.12), we merely shift the *p*-adic point one place to the right to obtain

$$(4.13) \qquad\qquad H(5, 4, \tfrac{2}{15}) = 4.131.$$

Case III

$$n > 0$$

In this case $\alpha = (c/d)p^{k}$, where k is a positive integer. To find $H(p, r, \alpha)$ we find $H(p, r, c/d)$ using the three steps in Case I, and then shift the *p*-adic point k places to the left.

For example, let $\alpha = \frac{10}{3}$, $p = 5$, and $r = 4$. We write $\alpha = (\frac{2}{3})5$ which gives us $c = 2$, $d = 3$, $k = 1$, and $p^r = 625$. Since we already have $H(5, 4, \frac{2}{3})$ in (4.12), we merely shift the *p*-adic point one place to the left to obtain

$$(4.14) \qquad\qquad H(5, 4, \tfrac{10}{3}) = .0413.$$

Notice that, since $r = 4$, we retain only four digits (including the zero) and the rightmost digit of $H(5, 4, \frac{2}{3})$ is discarded.

Hensel codes for negative rational numbers

In Theorem 2.19 we stated the relationship that exists between the (infinite) *p*-adic expansions for α and for $-\alpha$. In Example 2.20 we observed that

$$(4.15) \qquad \begin{cases} \tfrac{1}{3} = .2313131\ldots & (p = 5) \\ -\tfrac{1}{3} = .3131313\ldots & (p = 5). \end{cases}$$

Consequently, the corresponding Hensel codes (for $p = 5$ and $r = 4$) are

$$(4.16) \qquad \begin{cases} H(5, 4, \tfrac{1}{3}) = .2313 \\ H(5, 4, -\tfrac{1}{3}) = .3131. \end{cases}$$

Notice that the leftmost (non-zero) digit of the Hensel code for a positive rational number is complemented with respect to p and subsequent digits are complemented with respect to $p - 1$. Thus, from (4.12), (4.13), and (4.14) we can write

$$(4.17) \qquad \begin{cases} H(5, 4, -\tfrac{2}{3}) = .1313 \\ H(5, 4, -\tfrac{2}{15}) = 1.313 \\ H(5, 4, -\tfrac{10}{3}) = .0131. \end{cases}$$

4.18. Remark. It should be pointed out that the algorithm we used in obtaining (4.12), (4.13), and (4.14) can be used for negative fractions as well as for positive fractions. It can also be used in those cases for which $\gcd(a, b) \neq 1$ and $\gcd(a, p) \neq 1$.

For example, let $\alpha = -\frac{2}{4}$. If $p = 5$ and $r = 4$, $m = 625$. Hence,

$$|-\tfrac{2}{4}|_{625} = 312,$$

which we obtain using Algorithm 6.26, Chapter I, with the seed matrix

$$\begin{bmatrix} 625 & 0 \\ 4 & -2 \end{bmatrix}.$$

Since the radix-5 equivalent of this integer is

$$312_{\text{ten}} = 2222_{\text{five}},$$

we have the Hensel code

$$H(5, 4, -\tfrac{2}{4}) = .2222.$$

It is easy to verify that $-\frac{1}{2}$, $-\frac{3}{6}$, $-\frac{4}{8}$, etc., all have this same Hensel code. We point out that

$$H(5, 4, \tfrac{1}{2}) = .3222$$

and that .2222 and .3222 are complements of each other.

Similarly, if $\alpha = \frac{10}{3}$ (the example we used in Case III, above), then

$$|\tfrac{10}{3}|_{625} = 420,$$

which we obtain by using the seed matrix

$$\begin{bmatrix} 625 & 0 \\ 3 & 10 \end{bmatrix}$$

in Algorithm 6.26, Chapter I. Since the radix-5 equivalent of this integer is

$$420_{\text{ten}} = 3140_{\text{five}},$$

we obtain the Hensel code

$$H(5, 4, \tfrac{10}{3}) = .0413,$$

and this agrees with (4.14) above.

Floating-point Hensel codes

In discussing (4.14), in Case III above, we made the observation that $H(5, 4, \frac{10}{3})$ and $H(5, 4, \frac{2}{3})$ are related by a shift of the *p*-adic point and that one of the "significant digits" in the Hensel code for $\frac{2}{3}$ is discarded when we generate the Hensel code for $\frac{10}{3}$. This can be avoided if we introduce the concept of a normalized floating-point Hensel code.

4.19. Definition. Let $\alpha = a/b$ with $a/b = (c/d)p^n$ and

$$\gcd(c, d) = \gcd(c, p) = \gcd(d, p) = 1.$$

Then

$$\hat{H}(p, r, \alpha) = (m_\alpha, e_\alpha),$$

with

$$m_\alpha = H\left(p, r, \frac{c}{d}\right)$$

and

$$e_\alpha = n,$$

is the *normalized floating-point Hensel code* for α. We call m_α the *mantissa* and e_α the *exponent*.

4.20. EXAMPLES. From (4.12), (4.13), and (4.14) we obtain

$$\hat{H}(5, 4, \tfrac{2}{3}) = (.4131, 0)$$

$$\hat{H}(5, 4, \tfrac{2}{15}) = (.4131, -1)$$

$$\hat{H}(5, 4, \tfrac{10}{3}) = (.4131, 1),$$

and from (4.17) we obtain

$$\hat{H}(5, 4, -\tfrac{2}{3}) = (.1313, 0)$$

$$\hat{H}(5, 4, -\tfrac{2}{15}) = (.1313, -1)$$

$$\hat{H}(5, 4, -\tfrac{10}{3}) = (.1313, 1).$$

Notice that the digit in $H(5, 4, \tfrac{2}{3})$ which was discarded when we formed $H(5, 4, \tfrac{10}{3})$ in (4.14) has been recovered when we form the floating-point Hensel code $\tfrac{10}{3}$. This will be significant when we examine arithmetic using Hensel codes as operands.

4.21. Remark. In the normalized floating-point Hensel codes notice that the mantissa m_α is the ordinary Hensel code for c/d (in Definition 4.19). Also notice that the p-adic point in m_α is at the left of the string of r digits and that the leftmost digit (the one next to the p-adic point) is non-zero. In other words, c/d is a p-adic unit (see Definition 3.11) and m_α is a finite segment of the infinite p-adic expansion for that p-adic unit. Thus, the mantissas in the normalized floating-point Hensel codes play the same role in this finite system as the p-adic units play in the infinite system.

4.22. Table.   Ordinary Hensel Codes[a] $H(5, 4, a/b)$.

$\begin{smallmatrix} & a \\ b & \end{smallmatrix}$	1	2	3	4	5	6	7	8
1	.1000	.2000	.3000	.4000	.0100	.1100	.2100	.3100
2	.3222	.1000	.4222	.2000	.0322	.3000	.1322	.4000
3	.2313	.4131	.1000	.3313	.0231	.2000	.4313	.1231
4	.4333	.3222	.2111	.1000	.0433	.4222	.3111	.2000
5	1.000	2.000	3.000	4.000	.1000	1.100	2.100	3.100
6	.1404	.2313	.3222	.4131	.0140	.1000	.2404	.3313
7	.3302	.1214	.4021	.2423	.0330	.3142	.1000	.4302
8	.2414	.4333	.1303	.3222	.0241	.2111	.4030	.1000
9	.4201	.3012	.2313	.1124	.0420	.4131	.3432	.2243
10	3.222	1.000	4.222	2.000	.3222	3.000	1.322	4.000
11	.1332	.2120	.3403	.4240	.0133	.1411	.2204	.3041
12	.3424	.1404	.4333	.2313	.0342	.3222	.1202	.4131
13	.2034	.4014	.1143	.3123	.0203	.2232	.4212	.1341
14	.4101	.3302	.2013	.1214	.0410	.4021	.3222	.2423
15	2.313	4.131	1.000	3.313	.2313	2.000	4.313	1.231
16	.1234	.2414	.3104	.4333	.0123	.1303	.2042	.3222
17	.3043	.1132	.4121	.2210	.0304	.3342	.1431	.4420

$\begin{smallmatrix} & a \\ b & \end{smallmatrix}$	9	10	11	12	13	14	15	16	17
1	.4100	.0200	.1200	.2200	.3200	.4200	.0300	.1300	.2300
2	.2322	.0100	.3322	.1100	.4322	.2100	.0422	.3100	.1422
3	.3000	.0413	.2231	.4000	.1413	.3231	.0100	.2413	.4231
4	.1433	.0322	.4111	.3000	.2433	.1322	.0211	.4000	.3433
5	4.100	.2000	1.200	2.200	3.200	4.200	.3000	1.300	2.300
6	.4222	.0231	.1140	.2000	.3404	.4313	.0322	.1231	.2140
7	.2214	.0121	.3423	.1330	.4142	.2000	.0402	.3214	.1121
8	.3414	.0433	.2303	.4222	.1241	.3111	.0130	.2000	.4414
9	.1000	.0301	.4012	.3313	.2124	.1420	.0231	.4432	.3243
10	2.322	.1000	3.322	1.100	4.322	2.100	.4222	3.100	1.422
11	.4324	.0212	.1000	.2332	.3120	.4403	.0340	.1133	.2411
12	.2111	.0140	.3020	.1000	.4424	.2404	.0433	.3313	.1342
13	.3321	.0401	.2430	.4410	.1000	.3034	.0114	.2143	.4123
14	.1134	.0330	.4431	.3142	.2343	.1000	.0201	.4302	.3013
15	3.000	.4131	2.231	4.000	1.413	3.231	.1000	2.413	4.231
16	.4402	.0241	.1421	.2111	.3340	.4030	.0310	.1000	.2234
17	.2024	.0113	.3102	.1240	.4234	.2323	.0412	.3401	.1000

[a] See Krishnamurthy, Rao, and Subramanian [1975a], p. 70. Reproduced with permission.

Two rational numbers with the same Hensel code

Consider the two distinct rational numbers α and β and their canonical p-adic expansions

(4.23)
$$\begin{cases} \alpha = a_n p^n + a_{n+1} p^{n+1} + \cdots \\ \beta = b_n p^n + b_{n+1} p^{n+1} + \cdots . \end{cases}$$

Even though $\alpha \neq \beta$ it is still possible for the r leading coefficients of α to be identical with the r leading coefficients of β, in which case $H(p, r, \alpha)$ will equal $H(p, r, \beta)$. The following theorem gives a characterization of this situation.

4.24. Theorem. *Let* $\alpha, \beta \in \mathbb{Q}$. *Then* $H(p, r, \alpha) = H(p, r, \beta)$ *if and only if* p^r *divides* $\alpha - \beta$.

PROOF. Let $H(p, r, \alpha) = H(p, r, \beta)$. Then

$$\alpha = a_n p^n + \cdots + a_{n+r-1} p^{n+r-1} + a_{n+r} p^{n+r} + \cdots$$

and

$$\beta = a_n p^n + \cdots + a_{n+r-1} p^{n+r-1} + b_{n+r} p^{n+r} + \cdots$$

where n can be positive, negative, or zero. If we form $\alpha - \beta$, we obtain the expansion

$$\alpha - \beta = (a_{n+r} - b_{n+r}) p^{n+r} + (a_{n+r+1} - b_{n+r+1}) p^{n+r+1} + \cdots$$
$$= p^r [(a_{n+r} - b_{n+r}) p^n + (a_{n+r+1} - b_{n+r+1}) p^{n+1} + \cdots]$$

and p^r divides $\alpha - \beta$. We leave the proof of the converse to the reader. □

Notice that if $\alpha = a/b$ and $\beta = g/h$, with $\gcd(b, p) = \gcd(h, p) = 1$, then α and β can be represented by the integers $|a \cdot b^{-1}|_m$ and $|g \cdot h^{-1}|_m$, respectively, where $m = p^r$. In Remark 4.18 we pointed out that these two integers, when written in radix-p notation, give us the Hensel codes for α and β (when we reverse the order of the digits). Thus, it is easy to prove the following result.

4.25. Theorem. *Let* $\alpha = a/b$ *and* $\beta = g/h$, *with* $\gcd(b, p) = \gcd(h, p) = 1$ *and* $m = p^r$. *Then* $H(p, r, \alpha) = H(p, r, \beta)$ *if and only if*

$$|a \cdot b^{-1}|_m = |g \cdot h^{-1}|_m,$$

that is, if and only if

$$a \cdot b^{-1} \equiv g \cdot h^{-1} \qquad (\bmod \ p^r).$$

4.26. Corollary. *Let* $\alpha = a/b$ *and* $\beta = g/h$, *with* $\gcd(b, p) = \gcd(h, p) = 1$ *and* $m = p^r$. *Then* $H(p, r, \alpha) = H(p, r, \beta)$ *if and only if*

$$|ah|_m = |bg|_m,$$

that is, if and only if

$$ah \equiv bg \qquad (\mathrm{mod}\ p^r).$$

PROOF. In Theorem 4.25 multiply both sides of the equation and the congruence by bh and simplify.   □

The reader should compare this theorem and corollary with Theorem 5.6 and Corollary 5.7 in Chapter I.

4.27. EXAMPLE.* Let $p = 5, r = 4$, so that $p^r = 625$. Consider the two rational numbers $\alpha = \frac{10}{13}$ and $\beta = -\frac{35}{17}$. Obviously, 625 divides $\alpha - \beta = 625/221$. Observe also that

$$10 \cdot 13^{-1}(625) = 10 \cdot 577$$
$$= 5770$$

and

$$-35 \cdot 17^{-1}(625) = -35 \cdot 478$$
$$= -16730$$

and that

$$5770 \equiv -16730 \qquad (\mathrm{mod}\ 625).$$

Finally, observe that

$$10 \cdot 17 \equiv 13(-35) \qquad (\mathrm{mod}\ 625).$$

Consequently, from either Theorem 4.24, Theorem 4.25, or Corollary 4.26 we have

$$H(5, 4, \tfrac{10}{13}) = H(5, 4, -\tfrac{35}{17}).$$

Notice that (see Example 6.31, Chapter I)

$$\left|10 \cdot 13^{-1}\right|_{625} = \left|-35 \cdot 17^{-1}\right|_{625}$$
$$= 145,$$

and if we convert this decimal number to its radix-5 representation, we obtain

$$145_{\text{ten}} = 1040_{\text{five}}.$$

Consequently,

$$H(5, 4, \tfrac{10}{13}) = H(5, 4, -\tfrac{35}{17})$$
$$= .0401$$

and this value agrees with $H(5, 4, \tfrac{10}{13})$ in Table 4.22.

*This example is due to Ruth Ann Lewis [1979].

Obviously, any element of the generalized residue class* $\mathbb{Q}_{145}$, that is, any rational number a/b for which

(4.28) $|a \cdot b^{-1}|_{625} = 145,$

has the same Hensel code as $\frac{10}{13}$. However, as we saw in Section 5, Chapter I, if $\mathbb{Q}_k$ contains an order-N Farey fraction, it contains only one. In this example, $\alpha = \frac{10}{13}$ is the unique order-17 Farey fraction contained in $\mathbb{Q}_{145}$ and $-\frac{35}{17}$ is merely one of the (infinitely many) remaining elements of $\mathbb{Q}_{145}$.

4.29. Remark. It should be emphasized that the Hensel codes referred to in Theorem 4.24 and Theorem 4.25 are ordinary Hensel codes, not normalized floating-point Hensel codes. For example,

$$\hat{H}(5, 4, \tfrac{10}{13}) = (.4014, 1),$$

whereas

$$\hat{H}(5, 4, -\tfrac{35}{17}) = (.4013, 1),$$

and the mantissas differ in the fourth digit. However, it is clear that agreement between floating-point Hensel codes implies agreement between ordinary Hensel codes but the converse is not necessarily true.

Exercises II.4

1. Write down all the order-7 Farey fractions, i.e., the set $\mathbb{F}_7$. Hint: Use a two dimensional array.

2. If $p = 7$ and $r = 4$, what is N?

3. Compute $H(5, 4, \alpha)$ for
   (a) $\alpha = \frac{1}{7}$ (e) $\alpha = \frac{13}{15}$
   (b) $\alpha = \frac{2}{14}$ (f) $\alpha = -\frac{17}{3}$
   (c) $\alpha = \frac{17}{3}$ (g) $\alpha = -\frac{15}{7}$
   (d) $\alpha = \frac{15}{7}$ (h) $\alpha = -\frac{13}{15}$.
   Hint: Use Algorithm 6.26, Chapter I. Check your results with the entries in Table 4.22.

4. Compute $\hat{H}(5, 4, \alpha)$ for
   (a) $\alpha = \frac{2}{14}$ (d) $-\frac{15}{7}$
   (b) $\alpha = \frac{15}{7}$ (e) $\frac{26}{30}$
   (c) $\alpha = -\frac{13}{15}$ (f) $-\frac{2}{5}$.

5. In Example 4.27 we showed that $H(5, 4, \frac{10}{13}) = H(5, 4, -\frac{35}{17})$. Find another fraction $\alpha = a/b$ for which
   $$H(5, 4, \alpha) = H(5, 4, \tfrac{10}{13}).$$

6. Complete the proof of Theorem 4.24.

7. Prove Theorem 4.25.

* Recall (5.3), Chapter I.

5 Arithmetic Operations on Hensel Codes

In Section 3 we discussed arithmetic operations in the field of p-adic numbers $\mathbb{Q}_p$. We illustrated addition, subtraction, multiplication, and division using (infinite) p-adic expansions as operands. Since the Hensel codes are merely finite segments of the p-adic expansions it is not surprising to discover that we can do arithmetic using the Hensel codes as operands. In addition, it is not surprising to discover that the rules for the arithmetic of Hensel codes are essentially the same as the rules for arithmetic in $\mathbb{Q}_p$.

Throughout this section we shall assume that the operands are normalized floating-point Hensel codes with $p = 5$ and $r = 4$ unless it is stated otherwise. With $m = 625$ we have $N = 17$.

Addition

Consider the following numerical computation:

(5.1) $\alpha = \frac{2}{3} + \frac{1}{5}.$

The Hensel codes

(5.2) $\begin{cases} \hat{H}(5, 4, \frac{2}{3}) = (.4131, 0) \\ \hat{H}(5, 4, \frac{1}{5}) = (.1000, -1) \end{cases}$

can be used to carry out this addition. We simply line up the p-adic points and do radix-5 arithmetic proceeding from left to right. Since $r = 4$, we have

$$
\begin{array}{r}
.4131 \\
1.000 \\
\hline
1.413
\end{array}
$$

Hence, the sum we seek has .1413 as its mantissa and -1 as its exponent, that is,

(5.3) $\hat{H}(5, 4, \alpha) = (.1413, -1).$

Since $\frac{13}{15}$ is the order-17 Farey fraction with the property that

(5.4) $\hat{H}(5, 4, \frac{13}{15}) = (.1413, -1),$

we map* $(.1413, -1)$ onto

(5.5) $\alpha = \frac{13}{15},$

and this is the correct answer; that is, we do not have pseudo-overflow.

---

*The method for mapping Hensel codes onto rational numbers is discussed in Section 7.

Subtraction

Just as we did in $\mathbb{Q}_p$, we treat subtraction as "complemented addition" in the sense that the subtrahend is complemented and added to the minuend. For example, to carry out the computation

$$(5.6) \qquad \alpha = \tfrac{2}{3} - \tfrac{1}{5}$$

using Hensel codes, we need

$$(5.7) \qquad \hat{H}(5, 4, -\tfrac{1}{5}) = (.4444, -1).$$

Then, as before, we line up the p-adic points and do radix-5 addition proceeding from left to right. Since $r = 4$, we have

$$
\begin{array}{r}
.4131 \\
4.444 \\
\hline
4.313
\end{array}
$$

Hence, the answer we seek has .4313 as its mantissa and -1 as its exponent, that is,

$$(5.8) \qquad \hat{H}(5, 4, \alpha) = (.4313, -1).$$

Since $\tfrac{7}{15}$ is the order-17 Farey fraction with the property that

$$(5.9) \qquad \hat{H}(5, 4, \tfrac{7}{15}) = (.4313, -1),$$

we map $(.4313, -1)$ onto

$$(5.10) \qquad \alpha = \tfrac{7}{15},$$

and this is the correct answer; that is, we do not have pseudo-overflow.

Multiplication

Consider the following numerical computation:

$$(5.11) \qquad \alpha = \tfrac{1}{3} \cdot \tfrac{6}{5}.$$

We carry out this computation using the floating-point Hensel codes

$$(5.12) \qquad \begin{cases} \hat{H}(5, 4, \tfrac{1}{3}) = (.2313, 0) \\ \hat{H}(5, 4, \tfrac{6}{5}) = (.1100, -1). \end{cases}$$

The algorithm involves *multiplying* the mantissas and *adding* the exponents. Hence, using radix-5 arithmetic (proceeding from left to right), we obtain

$$
\begin{array}{r}
.2313 \\
.1100 \\
\hline
2313 \\
231 \\
\hline
.2000
\end{array}
$$

as the mantissa and $0 + (-1) = -1$ as the exponent for the product, that is,

(5.13) $\hat{H}(5, 4, \alpha) = (.2000, -1)$.

Since $\frac{2}{5}$ is the order-17 Farey fraction with the property that

(5.14) $\hat{H}(5, 4, \frac{2}{5}) = (.2000, -1)$,

we map $(.2000, -1)$ onto

(5.15) $\alpha = \frac{2}{5}$,

and this is the correct result; that is, we do not have pseudo-overflow.

Division

Consider the following numerical computation:

(5.16) $\alpha = \frac{2}{15} \div \frac{4}{15}$.

We carry out this computation using the floating-point Hensel codes

(5.17) $\begin{cases} \hat{H}(5, 4, \frac{2}{15}) = (.4131, -1) \\ \hat{H}(5, 4, \frac{4}{15}) = (.3313, -1). \end{cases}$

The algorithm involves *dividing* the mantissas and *subtracting* the exponents. We divide the mantissas the way we divided p-adic units in Section 3 (see Example 3.26). We need the multiplicative inverse, modulo p, of the leftmost digit of the divisor. In this example, we divide .4131 by .3313 and so we need

(5.18) $3^{-1}(5) = 2$.

We proceed exactly as we did in Example 3.26 except that we terminate as soon as we have r digits in the quotient. Hence,

$$
\begin{array}{r}
.3222 \\
.3313\overline{)\,.4131} \\
\underline{1444} \\
131 \\
421 \\
13 \\
\underline{42} \\
1 \\
\underline{4} \\
0
\end{array}
$$

and so the quotient we seek has .3222 for its mantissa. When we subtract exponents we get zero. Hence,

(5.19) $\hat{H}(5, 4, \alpha) = (.3222, 0)$.

Since $\frac{1}{2}$ is the order-17 Farey fraction with the property that

(5.20) $\hat{H}(5, 4, \frac{1}{2}) = (.3222, 0)$, we map $(.3222, 0)$ onto

(5.21) $$\alpha = \tfrac{1}{2},$$

and this is the correct result; that is, we do not have pseudo-overflow.

5.22. EXAMPLE. As a special case of the division in (5.16), let us consider the numerical computation

$$\beta = 1 \div \tfrac{4}{15}$$
$$= \tfrac{15}{4}.$$

In other words, if $\alpha = 4/15$, then $\beta = 1/\alpha$.

If we use the Hensel codes

$$\hat{H}(5, 4, 1) = (.1000, 0),$$
$$\hat{H}(5, 4, \tfrac{4}{15}) = (.3313, -1),$$

and if we divide mantissas and subtract exponents, we obtain

$$\hat{H}\left(5, 4, \frac{1}{\alpha}\right) = (.2111, 1).$$

Since $\tfrac{15}{4}$ is the order-17 Farey fraction with the property that

$$\hat{H}(5, 4, \tfrac{15}{4}) = (.2111, 1),$$

we map $(.2111, 1)$ onto

$$\frac{1}{\alpha} = \frac{15}{4},$$

and this is the correct reciprocal; that is, we do not have pseudo-overflow.

Computing the reciprocal of a Hensel code using Newton's method

One method for obtaining the mantissa .2111 in Example 5.22 is the "long division" method

$$
\begin{array}{r}
.2111 \\
.3313\overline{)\,.1000} \\
\underline{4213} \\
313 \\
213 \\
\underline{31} \\
21 \\
\underline{3} \\
2 \\
\underline{0}
\end{array}
$$

where the quotient is obtained digit-by-digit. However, it is possible to accelerate this sequential process by using a generalization of Newton's method for finding the reciprocal of a real number. See Krishnamurthy [1970], [1971], and Krishnamurthy and Murthy [1983].

Recall that we use Newton's method to compute a root of the equation $f(x) = 0$ by generating a sequence of approximations $x_0, x_1, x_2, \ldots$ using the iteration formula

$$(5.23) \qquad x_{i+1} = x_i - \frac{f(x_i)}{f'(x_i)}, \qquad i = 0, 1, 2, \ldots,$$

where x_0 is a "suitably chosen" initial approximation to the root. If we are given the function

$$(5.24) \qquad f(x) = \frac{1}{x} - a,$$

then the iteration function (when simplified) becomes

$$(5.25) \qquad x_{i+1} = x_i(2 - ax_i), \qquad i = 0, 1, 2, \ldots$$

and the sequence of approximations produced by (5.25) converges to the reciprocal of a. Because Newton's method has quadratic convergence (for simple roots), if x_0 contains at least one significant digit of accuracy, the number of significant digits doubles during each step of the iteration.

In order to generalize Newton's method to handle p-adic numbers (and Hensel codes) we need to guarantee that

(i) Newton's iteration is applicable and valid from the point of view of convergence in the p-adic field, and

(ii) the number of digits in the mantissa of a Hensel code is doubled at each step of the iteration.

Since we only use the mantissa of the normalized floating-point Hensel code in Example 5.22, and since (ii) above only refers to the number of digits in the mantissa, there is no loss in generality if we restrict further discussion to those numbers α for which the exponent is zero. In this case the mantissa of α is its ordinary Hensel code. Therefore, in (ii) above we are guaranteeing that

$$H(p, s, \alpha) \rightarrow H(p, 2s, \alpha).$$

Hensel's lemma (see, for example, Koblitz [1977], page 16) guarantees (i) and Theorem 5.36, below, guarantees (ii).

For convenience, we write

$$(5.26) \qquad H(p, r, \alpha) = .a_0 a_1 \ldots a_{r-1}$$

which corresponds to the integer

$$(5.27) \qquad a = a_0 + a_1 p + a_2 p^2 + \cdots + a_{r-1} p^{r-1}$$

(with $a_0 \neq 0$) and

$$(5.28) \qquad H\left(p, r, \frac{1}{\alpha}\right) = .c_0 c_1 \ldots c_{r-1}$$

which corresponds to the integer

$$(5.29) \qquad c = c_0 + c_1 p + c_2 p^2 + \cdots + c_{r-1} p^{r-1}$$

(with $c_0 \neq 0$). Observe that

$$(5.30) \qquad |a \cdot c|_{p^r} = 1,$$

which implies $c = a^{-1}(p^r)$, that is, c is the multiplicative inverse of a modulo p^r. Also, observe that the product of the two Hensel codes is $.100 \ldots 0$, which implies

$$(5.31) \qquad c_0 = a_0^{-1}(p).$$

Suppose we introduce the notation

$$(5.32) \quad \begin{cases} b_1 = c_0 \\[4pt] b_2 = c_0 + c_1 p \\[4pt] b_4 = c_0 + c_1 p + c_2 p^2 + c_3 p^3 \\[4pt] \quad \vdots \\[4pt] b_r = c_0 + c_1 p + \cdots + c_{r-1} p^{r-1} \end{cases} \quad \begin{cases} H\left(p, 1, \frac{1}{\alpha}\right) = .c_0 \\[4pt] H\left(p, 2, \frac{1}{\alpha}\right) = .c_0 c_1 \\[4pt] H\left(p, 4, \frac{1}{\alpha}\right) = .c_0 c_1 c_2 c_3 \\[4pt] \quad \vdots \\[4pt] H\left(p, r, \frac{1}{\alpha}\right) = .c_0 c_1 \ldots c_{r-1} \end{cases}$$

where $r = 2^i$. Notice that $H(p, r, 2) = .200 \cdots 0$ and recall (5.26) and (5.27). Then Newton's iteration allows us to write

$$(5.33) \qquad b_{2k} = \left| b_{2k-1}(2 - a b_{2k-1}) \right|_{p^{2k}}, \qquad k = 1, 2, \ldots, i,$$

or equivalently,

$$(5.34) \qquad \begin{aligned} H\left(p, 2^k, \frac{1}{\alpha}\right) &= H\left(p, 2^{k-1}, \frac{1}{\alpha}\right) \\[6pt] &\quad * \left[H(p, 2^k, 2) - H(p, 2^k, \alpha) * H\left(p, 2^{k-1}, \frac{1}{\alpha}\right) \right] \end{aligned}$$

where $*$ denotes the multiplication of Hensel codes, and the "missing" digits in $H\left(p, 2^{k-1}, \frac{1}{\alpha}\right)$ are filled in with zeros.

In Theorem 5.36 we justify (5.33). However, before we prove this theorem, we illustrate the algorithm with a computation.

5.35. EXAMPLE. Suppose we are given $H(5, 4, \frac{3}{11}) = .3403$ and we wish to find $H(5, 4, \frac{11}{3})$. We proceed as follows. First, since $a_0 = 3$ and $c_0 = a_0^{-1}(5)$, we find $c_0 = 2$. Therefore,

$$H(5, 1, \tfrac{11}{3}) = .2$$

is our single-digit approximation. Next, we compute

$$H(5, 2, \tfrac{11}{3}) = (.20) * [(.20) - (.34) * (.20)]$$

$$= (.20) * [(.20) - (.14)]$$

$$= (.20) * (.11)$$

$$= .22$$

and we have our two-digit approximation. Finally, we compute

$$H(5, 4, \tfrac{11}{3}) = (.2200) * [(.2000) - (.3403) * (.2200)]$$

$$= (.2200) * [(.2000) - (.1013)]$$

$$= (.2200) * (.1041)$$

$$= .2231$$

and we have the final result.

5.36. **Theorem.** *There exists a sequence of integers*

$$\{b_1, b_2, b_4, \ldots, b_r, \ldots\}$$

with $r = 2^i$, such that

$$|a \cdot b_{2k}|_{p^{2k}} = 1, \qquad k = 0, 1, 2, \ldots, i, \ldots$$

where a, defined in (5.27), corresponds to the Hensel code $H(p, r, \alpha)$ defined in (5.26), and b_{2k} corresponds to the Hensel code $H\left(p, 2^k, \dfrac{1}{\alpha}\right)$ defined in (5.32).

PROOF. (by induction). Since we have defined b_1 to be $a_0^{-1}(p)$, we have*

$$|a \cdot b_1|_p = |(a_0 + a_1 p + \cdots + a_{r-1} p^{r-1}) a_0^{-1}|_p$$

$$= |a_0 \cdot a_0^{-1}|_p$$

$$= 1.$$

Now assume (inductive hypothesis) that

$$|a \cdot b_{2k-1}|_{p^{2k-1}} = 1.$$

Then,

$$|a \cdot b_{2k}|_{p^{2k}} = |a|(2 - a \cdot b_{2k-1}) b_{2k-1}|_{p^{2k}}|_{p^{2k}}$$

$$= |a \cdot b_{2k-1}(2 - a \cdot b_{2k-1})|_{p^{2k}}$$

and if we use the fact that our inductive hypothesis implies

---

* We simplify $a_0^{-1}(p)$ to a_0^{-1}, whenever p is understood.

$$a \cdot b_{2k-1} = 1 + m \cdot p^{2^{k-1}}$$

for some integer m, we are able to write

$$\begin{aligned}
|a \cdot b_{2k}|_{p^{2k}} &= |(1 + m \cdot p^{2^{k-1}})(2 - [1 + m \cdot p^{2^{k-1}}])|_{p^{2k}} \\
&= |(1 + m \cdot p^{2^{k-1}})(1 - m \cdot p^{2^{k-1}})|_{p^{2k}} \\
&= 1.
\end{aligned}$$

Since, by construction, the theorem holds for $k = 0$, and since the assumption that the result is true for $k - 1$ implies that it is true for k, it is true for all $k \geq 0$. $\square$

Higher order convergence

In the iterative procedure above, we obtain quadratic convergence in the sense that the number of p-adic digits is doubled at every step. Krishnamurthy [1971] describes a procedure for extending this to cubic and higher orders of convergence.

To describe the extension to higher orders of convergence we first write (5.33) in the form

(5.37)
$$\begin{aligned}
b_{2k} &= |b_{2k-1}[1 + (1 - a \cdot b_{2k-1})]|_{p^{2k}} \\
&= |b_{2k-1}[1 + d_{k-1}]|_{p^{2k}}, \qquad k = 1, 2, \ldots, i,
\end{aligned}$$

where

(5.38a)
$$d_{k-1} = 1 - a \cdot b_{2k-1}.$$

To obtain convergence of order $q > 2$, we replace the 2 in (5.38a) by q, that is, we write

(5.38b)
$$d_{k-1} = 1 - a \cdot b_{qk-1}.$$

Then we iterate, using the equation

(5.39)
$$b_{qk} = |b_{qk-1}[1 + d_{k-1}(1 + d_{k-1}(1 + \ldots) \ldots)]|_{p^{qk}},$$

where there are $q - 1$ terms in the nested expression. For example, for cubic convergence, we use

(5.40)
$$\begin{aligned}
b_{3k} &= |b_{3k-1}[1 + d_{k-1}(1 + d_{k-1})]|_{p^{3k}} \\
&= |b_{3k-1}[1 + (1 - a \cdot b_{3k-1})(2 - a \cdot b_{3k-1})]|_{p^{3k}}.
\end{aligned}$$

5.41. EXAMPLE. Suppose we are given $H(5, 9, 7) = .210000000$ and we wish to find $H(5, 9, \frac{1}{7})$. We proceed as follows. First, since $a_0 = 2$, we compute $c_0 = 3$, and so

$$H(5, 1, \tfrac{1}{7}) = .3$$

is our single-digit approximation. Next, we compute

$$H(5, 3, \tfrac{1}{7}) = (.300) * [(.100) + \{(.100) - (.210) * (.300)\}$$
$$* \{(.200) - (.210) * (.300)\}]$$
$$= (.300) * [(.100) + \{(.100) - (.140)\}$$
$$* \{(.200) - (.140)\}]$$
$$= (.300) * [(.100) + (.014) * (.114)]$$
$$= (.300) * [(.100) + (.010)]$$
$$= (.300) * (.110)$$
$$= .330$$

and we have our three-digit approximation. Finally, we compute

$$H(5, 9, \tfrac{1}{7}) = (.330000000) * [(.100000000)$$
$$+ \{(.100000000) - (.210000000) * (.330000000)\}$$
$$* \{(.200000000) - (.210000000) * (.330000000)\}]$$
$$= (.330000000) * [(.100000000)$$
$$+ \{(.100000000) - (.100100000)\}$$
$$* \{(.200000000) - (.100100000)\}]$$
$$= (.330000000) * [(.100000000)$$
$$+ \{.000444444\} * \{.100444444\}]$$
$$= (.330000000) * [(.100000000) + (.000444000)]$$
$$= (.330000000) * (.100444000)$$
$$= .330214230$$

and we have the final result.

EXERCISES II.5

1. Use finite-segment *p*-adic arithmetic with $p = 5$ and $r = 4$.
   - (a) Add $\tfrac{4}{3}$ to $\tfrac{1}{6}$.
   - (b) Add $\tfrac{1}{3}$ to $\tfrac{5}{2}$.
   - (c) Subtract $\tfrac{4}{3}$ from $\tfrac{1}{6}$.
   - (d) Subtract $\tfrac{1}{3}$ from $\tfrac{5}{2}$.
   - (e) Multiply $\tfrac{4}{3}$ by $\tfrac{1}{6}$.
   - (f) Multiply $\tfrac{1}{3}$ by $\tfrac{5}{2}$.
   - (g) Divide $\tfrac{4}{3}$ by $\tfrac{1}{6}$.
   - (h) Divide $\tfrac{1}{3}$ by $\tfrac{5}{2}$.

2. First find the Hensel code for α. Then find
   - (a) $H(7, 8, 1/\alpha)$, where $\alpha = 3/4$, using Newton's quadratic method,
   - (b) $H(7, 9, 1/\alpha)$, where $\alpha = 3/4$, using Newton's cubic convergent method,
   - (c) $H(101, 8, 1/\alpha)$, where $\alpha = 2/7$, using Newton's quadratic method, and
   - (d) $H(601, 9, 1/\alpha)$, where $\alpha = 7/11$, using Newton's cubic convergent method.

6 Removing a Leading Zero from a Hensel Code

In Definition 4.19 we assumed that

$$(6.1) \qquad \alpha = \left(\frac{c}{d}\right) p^n$$

with $\gcd(c, d) = \gcd(c, p) = \gcd(d, p) = 1$, and this enables us to write the normalized floating-point Hensel code in the form

$$(6.2) \qquad \hat{H}(p, r, \alpha) = (m_\alpha, e_\alpha),$$

where

$$(6.3) \qquad m_\alpha = H\left(p, r, \frac{c}{d}\right).$$

In other words, the mantissa in the floating-point Hensel code is the ordinary Hensel code for c/d. In this case, of course, the leftmost digit of m_α is different from zero which means m_α can be used as the divisor in a division operation.

There are occasions following the operations of addition and/or subtraction when the mantissa is not normalized, that is, the mantissa contains a zero in the leftmost digit position, and yet the mantissa is the divisor in a division operation. Since the leftmost digit in the Hensel code of a divisor must have a multiplicative inverse modulo p, this creates some difficulty.

6.4. EXAMPLE. Suppose we wish to compute

$$x = \frac{a}{b + c}$$

with $b = \frac{1}{2}$ and $c = \frac{1}{8}$. If we use the floating-point Hensel codes

$$\hat{H}(5, 4, \tfrac{1}{2}) = (.3222, 0)$$

and

$$\hat{H}(5, 4, \tfrac{1}{8}) = (.2414, 0),$$

we obtain the mantissa of the floating-point Hensel code for $y = b + c$ by adding the mantissas, as follows:

$$
\begin{array}{r}
.3222 \\
.2414 \\
\hline
.0241
\end{array}
$$

Hence, the *unnormalized* floating-point Hensel code of y is $(.0241, 0)$.

Observe that this unnormalized floating-point Hensel code cannot be used as the divisor in a division operation because of the leading zero in the mantissa. (See Section 5.) Consequently, we need to normalize it.

An obvious procedure for normalizing the Hensel code involves two steps. First, we map the unnormalized Hensel code onto the order-N Farey fraction y, and then we map y onto its normalized floating-point Hensel code. For example, by using the method described in the next section, we find that $(.0241, 0)$ maps onto

$$y = \tfrac{5}{8}$$

$$= (\tfrac{1}{8})5.$$

Consequently, since $\hat{H}(5, 4, \tfrac{1}{8})$ is given above, we can map $\tfrac{5}{8}$ onto the normalized floating-point Hensel code

$$\hat{H}(5, 4, \tfrac{5}{8}) = (.2414, 1).$$

6.5. Remark. Notice that the unnormalized floating-point Hensel code $(.0241, 0)$, whose mantissa contains a leading zero, cannot be normalized simply by shifting the mantissa left and adjusting the exponent, because we do not know the fourth digit in

$$(.241x, 1).$$

Consequently, we must use the method described in Example 6.4 to find that the unnormalized Hensel code $(.0241, 0)$ and the normalized Hensel code $(.2414, 1)$ are *equivalent*, that is, that they correspond to the same order-N Farey fraction. See Remark 7.2 in the next section.

7 Mapping a Hensel Code onto a Unique Order-N Farey Fraction

In the last section we showed that if we could map an unnormalized mantissa onto a unique order-N Farey fraction, then we could normalize an unnormalized floating-point Hensel code. In this section our objective is to show how this mapping can be accomplished. (It turns out that it doesn't matter whether the mantissa is normalized or not.)

The algorithm involves the following steps:

(i) Map the mantissa onto a radix-p integer by reversing the order of the digits.
(ii) Map the radix-p integer onto a radix-β integer, where β is the radix of the number system we wish to use for our computation. (We will use $\beta = 10$ in our examples.)
(iii) Use the methods of Section 6, Chapter I, to map the integer in $\hat{I}_m$ (where $m = p^r$) onto a unique order-N Farey fraction.

7.1. EXAMPLE. In Example 6.4 we mentioned that the unnormalized floating-point Hensel code $(.0241, 0)$ maps onto $y = \tfrac{5}{8}$. (Recall that $p = 5$, $r = 4$, $m = 625$, and $N = 17$.) This value of y is obtained as follows:

(i) The unnormalized mantissa is mapped onto the radix-5 integer

$$.0241 \rightarrow 1420_{\text{five}}.$$

(ii) The radix-5 integer is mapped onto the radix-10 integer

$$1420_{\text{five}} = 235_{\text{ten}}.$$

(iii) We can use either Algorithm 6.26, Chapter I, or the "common denominator" method to map 235 onto the order-17 Farey fraction $\frac{5}{8}$.

Using Algorithm 6.26 we obtain the table

	625	0
	235	1
2	155	-2
1	80	3
1	75	-5
1	5	8
15	0	-125

from which we observe that

$$y = \tfrac{5}{8}.$$

Using the "common denominator" method requires that we find a common denominator when we form the sum

$$y = \tfrac{1}{2} + \tfrac{1}{8}.$$

Obviously, the sum is a rational number u/v and we can see that a common denominator is $2 \cdot 8 = 16$. Hence, we let $tv = 16$. From (ii),

$$k = \left| u \cdot v^{-1} \right|_{625}$$

$$= 235,$$

and so

$$tu = /(tv)k/_{625}$$

$$= /16 \cdot 235/_{625}$$

$$= 10.$$

Thus, again we obtain

$$y = \tfrac{5}{8}.$$

7.2. Remark. Once we have mapped the mantissa of a floating-point Hensel code (either normalized or unnormalized) onto an order-N Farey fraction, there remains the task of multiplying this order-N Farey fraction by p^n, where n is the exponent in the floating-point Hensel code. For example, in

Remark 6.5 we stated that $(.0241, 0)$ and $(.2414, 1)$ are equivalent. We now demonstrate that this is true.

We have just shown that the unnormalized mantissa $.0241$ maps onto the order-17 Farey fraction $y = \frac{5}{8}$. Since the corresponding exponent is zero,

$$(.0241, 0) \to \tfrac{5}{8}.$$

It is easy to see that the normalized mantissa $.2414$ maps onto the order-17 Farey fraction $\tilde{y} = \frac{1}{8}$, because

$$.2414 \to 4142_{\text{five}} = 547_{\text{ten}},$$

and if we use Algorithm 6.26, Chapter I, we obtain

		625	0
		547	1
1		78	-1
7		1	8
78		0	-625

Since the exponent is $n = 1$, $y = 5\tilde{y}$, and so

$$(.2414, 1) \to \tfrac{5}{8}.$$

This demonstrates that both the normalized and the unnormalized Hensel codes represent the same rational number.

7.3. EXAMPLE. Suppose

$$\alpha = \frac{u}{v}$$

$$= \left(\frac{c}{d}\right) p^n,$$

where $\gcd(c, d) = \gcd(c, p) = \gcd(d, p) = 1$, with the ordinary Hensel code

$$H(5, 4, \alpha) = 3.222,$$

and the normalized floating-point Hensel code

$$\hat{H}(5, 4, \alpha) = (.3222, -1).$$

The mantissa $.3222$ maps onto the integer

$$2223_{\text{five}} = 313_{\text{ten}},$$

which implies

$$|c \cdot d^{-1}|_{625} = 313.$$

(a) If we use Algorithm 6.26, Chapter I, we obtain

$$
\begin{array}{c|cc}
 & 625 & 0 \\
 & 313 & 1 \\
1 & 312 & -1 \\
1 & 1 & 2 \\
\hline
312 & 0 & -625
\end{array}
$$

which implies

$$\frac{c}{d} = \frac{1}{2}.$$

Finally, since the exponent is $n = -1$, we must multiply c/d by $1/5$ to obtain α, that is,

$$\alpha = \tfrac{1}{10}.$$

Observe that this agrees with what we find in Table 4.22.

(b) Suppose we happen to know that $gv = 20$. Since an integral multiple of v equals some integral multiple of d, say td, we have $td = 20$. Hence, we can compute tc using the "common denominator" method. Therefore,

$$tc = /20 \cdot 313/_{625}$$

$$= 10,$$

which implies

$$\frac{c}{d} = \frac{1}{2},$$

and, as in (a),

$$\alpha = \tfrac{1}{10}.$$

Notice that the "common denominator" method used in part (b) is based on the assumption that we somehow know an integral multiple of v, say gv, where $0 < g \leq N$. Unfortunately, we have no way of determining, *a priori*, whether or not g lies in this range. All we can do is use the algorithm and then test our value of u/v to see whether it is an order-N Farey fraction or one of the other rational numbers with the same Hensel code as the unique rational number we seek.

Suppose we know the sequence of arithmetic operations which produces the rational number u/v. One approach to finding gv is to maintain a "running denominator" as the computation progresses so that, once the computation is finished, the "running denominator" is our gv.

To be more specific, when the Hensel codes for the initial data are determined, the denominator of each rational operand for addition, subtraction and multiplication (and the numerator of each rational divisor) can be stored with its Hensel code. Then, following each arithmetic operation on a pair of Hensel codes, the Hensel code for the running denominator can be computed and stored with the Hensel code resulting from the operation. For example following an addition or subtraction operation on a pair of Hensel codes, the least common multiple of the denominators associated with the two operands constitutes a running denominator. For a detailed discussion of this procedure see Rao [1975], pp. 149–151, and Krishnamurthy et al. [1975b], pp. 170–171.

A better procedure can be used when solving certain problems by virtue of the fact that a mathematical property of a problem may provide us with a simple algorithm for finding a common denominator. For example, if $A = (a_{ij})$ is a nonsingular matrix whose elements a_{ij} are integers, then the inverse matrix $A^{-1} = (\det A)^{-1} A^{adj}$ has the property that the adjoint matrix A^{adj} has elements which are integers. Consequently, the integer $\det A$ is a common denominator for the rational elements of A^{-1}.

If we solve the system of linear algebraic equations $Ax = b$, then

(7.4)
$$x = A^{-1}b$$
$$= (\det A)^{-1} A^{adj} b.$$

Thus, if the components of the vector b are rational numbers with a common denominator d, then x has rational elements with a common denominator $d(\det A)$. If Gaussian elimination is used, $\det A$ equals the product of the pivotal elements and is computed as a by-product of the computation.

7.5. EXAMPLE. Let us solve the system of linear algebraic equations $Ax = b$ where

$$A = \begin{bmatrix} 2 & 2 & -1 \\ -3 & 0 & 2 \\ 4 & -5 & -1 \end{bmatrix} \quad \text{and} \quad b = \begin{bmatrix} 3 \\ -\frac{7}{2} \\ \frac{1}{2} \end{bmatrix}.$$

If we use Gaussian elimination (with scaling), we obtain

$$\begin{bmatrix} 2 & 2 & -1 & 3 \\ -3 & 0 & 2 & -\frac{7}{2} \\ 4 & -5 & -1 & \frac{1}{2} \end{bmatrix} \rightarrow \begin{bmatrix} 1 & 1 & -\frac{1}{2} & \frac{3}{2} \\ -3 & 0 & 2 & -\frac{7}{2} \\ 4 & -5 & -1 & \frac{1}{2} \end{bmatrix}$$

$$\rightarrow \begin{bmatrix} 1 & 1 & -\frac{1}{2} & \frac{3}{2} \\ 0 & 3 & \frac{1}{2} & 1 \\ 0 & -9 & 1 & -\frac{11}{2} \end{bmatrix} \rightarrow \begin{bmatrix} 1 & 1 & -\frac{1}{2} & \frac{3}{2} \\ 0 & 1 & \frac{1}{6} & \frac{1}{3} \\ 0 & -9 & 1 & -\frac{11}{2} \end{bmatrix}$$

$$\rightarrow \begin{bmatrix} 1 & 1 & -\frac{1}{2} & \frac{3}{2} \\ 0 & 1 & \frac{1}{6} & \frac{1}{3} \\ 0 & 0 & \frac{5}{2} & -\frac{5}{2} \end{bmatrix} \rightarrow \begin{bmatrix} 1 & 1 & -\frac{1}{2} & \frac{3}{2} \\ 0 & 1 & \frac{1}{6} & \frac{1}{3} \\ 0 & 0 & 1 & -1 \end{bmatrix}$$

which represents the triangular system

$$\begin{bmatrix} 1 & 1 & -\frac{1}{2} \\ 0 & 1 & \frac{1}{6} \\ 0 & 0 & 1 \end{bmatrix} \begin{bmatrix} x_1 \\ x_2 \\ x_3 \end{bmatrix} = \begin{bmatrix} \frac{3}{2} \\ \frac{1}{3} \\ -1 \end{bmatrix}.$$

Observe that the only division operations in this algorithm involved scaling the rows by the three pivots 2, 3, and $\frac{5}{2}$. Observe, also, that

$$\det A = 15,$$

the product of the pivots. Now, if we use back substitution, we obtain

$$\begin{bmatrix} x_1 \\ x_2 \\ x_3 \end{bmatrix} = \begin{bmatrix} \frac{1}{2} \\ \frac{1}{2} \\ -1 \end{bmatrix}.$$

Suppose we use finite-segment *p*-adic arithmetic for this computation. It can be shown (for example, see Young and Gregory [1973], p. 883) that it is sufficient to select p and r in such a way that

$$p^r \geq 2 \left[\sum_{j=1}^{n} |b_j| \right] \prod_{i=1}^{n} \left[\sum_{k=1}^{n} a_{ik}^2 \right]^{1/2}.$$

However, this condition is sufficient, but not necessary, and is often a very conservative bound for p^r. It turns out that if we select $p = 5$ and $r = 4$, we can handle the example above without pseudo-overflow, even though $p^r = 625$ and the right-hand side of the inequality is approximately 981.

We begin by writing the augmented matrix $[A, b]$ in *p*-adic form using normalized floating-point Hensel codes.* This gives us

$$[A, b] = \begin{bmatrix} (.2000, 0) & (.2000, 0) & (.4444, 0) & (.3000, 0) \\ (.2444, 0) & (.0000, 0) & (.2000, 0) & (.4122, 0) \\ (.4000, 0) & (.4444, 1) & (.4444, 0) & (.3222, 0) \end{bmatrix}$$

Using Gaussian elimination (with scaling) we obtain the matrices

$$\begin{bmatrix} (.1000, 0) & (.1000, 0) & (.2222, 0) & (.4222, 0) \\ (.0000, 0) & (.3000, 0) & (.3222, 0) & (.1000, 0) \\ (.0000, 0) & (.1344, 0) & (.1000, 0) & (.2122, 0) \end{bmatrix},$$

$$\begin{bmatrix} (.1000, 0) & (.1000, 0) & (.2222, 0) & (.4222, 0) \\ (.0000, 0) & (.1000, 0) & (.1404, 0) & (.2313, 0) \\ (.0000, 0) & (.0000, 0) & (.3222, 1) & (.2222, 1) \end{bmatrix},$$

and

$$\begin{bmatrix} (.1000, 0) & (.1000, 0) & (.2222, 0) & (.4222, 0) \\ (.0000, 0) & (.1000, 0) & (.1404, 0) & (.2313, 0) \\ (.0000, 0) & (.0000, 0) & (.1000, 0) & (.4444, 0) \end{bmatrix}.$$

*The normalized floating-point Hensel code for zero is (.0000, 0).

The product of the pivots gives us

$$\hat{H}(5, 4, \det A) = (.2000, 0) \cdot (.3000, 0) \cdot (.3222, 1)$$
$$= (.3000, 1),$$

and so $\det A = 15$.

To solve the triangular system of equations we use back substitution. Thus, we have the following Hensel codes:

$$\hat{H}(5, 4, x_3) = (.4444, 0),$$
$$\hat{H}(5, 4, x_2) = (.2313, 0) - (.4444, 0) \cdot (.1404, 0)$$
$$= (.3222, 0),$$

and

$$\hat{H}(5, 4, x_1) = (.4222, 0) - (.4444, 0) \cdot (.2222, 0) - (.3222, 0) \cdot (.1000, 0)$$
$$= (.3222, 0).$$

This implies

$$|x_3|_{625} = 624$$
$$|x_2|_{625} = 313$$

and

$$|x_1|_{625} = 313.$$

Since $b = [3, -\frac{7}{2}, \frac{1}{2}]^T$, these components have a common denominator $d = 2$. Thus, our common denominator for the components of x is

$$d(\det A) = 30.$$

Hence, we have

$$x_3 = \tfrac{1}{30}/30 \cdot 624/_{625}$$
$$x_2 = \tfrac{1}{30}/30 \cdot 313/_{625}$$

and

$$x_1 = \tfrac{1}{30}/30 \cdot 313/_{625},$$

which imply

$$\begin{bmatrix} x_1 \\ x_2 \\ x_3 \end{bmatrix} = \begin{bmatrix} \frac{1}{2} \\ \frac{1}{2} \\ -1 \end{bmatrix}.$$

It is easily verified that Algorithm 6.26, Chapter I, can be used to produce these same results.

7.6. Remark. It is shown in Krishnamurthy et al. [1975a] that if r is even and $\alpha \in \mathbb{F}_N$ is a *positive integer*, then the last $r/2$ digits in $H(p, r, \alpha)$ are all zero. For example, if $p = 5$, $r = 8$, and $\alpha = 199$,

$$H(5, 8, 199) = .4421\,0000.$$

Similarly, if α is a *negative integer*, the last $r/2$ digits in $H(p, r, \alpha)$ are all $p - 1$. (Recall the complement scheme for representing negative numbers.) For example,

$$H(5, 8, -199) = .1023\,4444.$$

When we have the Hensel code for a rational number u/v (not an integer) we can add it to itself (using v summands) and produce the Hensel code for the integer u because of the simple fact that

$$v(u/v) = u.$$

Obviously, since $|v| \leq N$, the number of summands is bounded by N. For example, let $H(5, 4, u/v) = .4131$. If we form the sum

$$
\begin{array}{r}
.4131 \\
.4131 \\
.4131 \\
\hline
.2000
\end{array}
$$

we observe that the sum represents the integer $u = 2$ and there are $v = 3$ summands. Hence, $u/v = 2/3$.

We include this method because of its simplicity, even though it is not as useful as the two methods previously discussed. For example, Beiser [1979] has discovered that it is possible to encounter a Hensel code which represents an integer by using fewer than v summands; in other words, we can "stumble onto" rational numbers in the same generalized residue class as u/v before we arrive at u/v.

A fourth method, based on a direct table look-up is described in Rao and Gregory [1981]. However, it is not a practical computational method and it will not be described here. It should be mentioned, however, that the reference cited contains some tables which shed some light on the mapping between the order-N Farey fractions and their integer representations.

EXERCISES II.7

1. Which normalized floating-point Hensel codes are equivalent to the following, where $p = 5$ and $r = 4$?
   (i) (.0121, 0)   (iii) (.0140, 0)
   (ii) (.0231, 0)   (iv) (.0330, 0).

2. Which normalized floating-point Hensel codes are equivalent to the following ordinary Hensel codes?

    (i) $H(5, 4, \alpha) = 4.200$ (iii) $H(5, 4, \alpha) = 2.322$

    (ii) $H(5, 4, \alpha) = 3.231$ (iv) $H(5, 4, \alpha) = 3.100$.

3. Find the order-17 Farey fraction which corresponds to each Hensel code in Problem 1.

4. Find the order-17 Farey fraction which corresponds to each Hensel code in Problem 2.

5. Find the order-6 Farey fraction α in each of the following:

    (i) $\hat{H}(3, 4, \alpha) = (.2101, 0)$

    (ii) $\hat{H}(3, 4, \alpha) = (.1202, 1)$

    (iii) $\hat{H}(3, 4, \alpha) = (.1111, -1)$.

6. Solve the system of linear algebraic equations $Ax = b$ with

$$A = \begin{bmatrix} 2 & -1 & -3 \\ 3 & 0 & 1 \\ -1 & 2 & 2 \end{bmatrix}, \quad b = \begin{bmatrix} \frac{1}{2} \\ 2 \\ -\frac{3}{2} \end{bmatrix},$$

using finite-segment p-adic arithmetic. Show that it is safe to use $p = 5$ and $r = 4$, that is, show that the inequality in Example 7.5 is satisfied.

CHAPTER III

Exact Computation of Generalized Inverses

1 Introduction

It is well known that every nonsingular real or complex (square) matrix A has a unique *inverse* which has the property that

$$(1.1) \qquad AA^{-1} = A^{-1}A = I.$$

This guarantees that the system of linear algebraic equations $Ax = b$ has the unique solution

$$(1.2) \qquad x = A^{-1}b.$$

A matrix has an inverse only if it is square, and a square matrix A has an inverse if and only if it is nonsingular, that is, if and only if

(i) det $A \neq 0$, or
(ii) the columns of A are linearly independent, or
(iii) the rows of A are linearly independent,

where each of these three properties implies the other two.

If a matrix is rectangular or if a square matrix is singular, it does not have an inverse, but it does have a *generalized inverse*, called a *g-inverse*, which has the following properties:

(i) a *g*-inverse exists for a class of matrices larger than the class of non-singular matrices,
(ii) a *g*-inverse has some of the properties of the ordinary matrix inverse, and

(iii) a g-inverse reduces to the ordinary matrix inverse, if A is square and nonsingular.

If A is an $m \times n$ matrix and G is a g-inverse of A, then G is an $n \times m$ matrix defined as follows.

1.3. **Definition.** Consider the matrix equations

(a) $$AGA = A$$

(b) $$GAG = G$$

(c) $$(AG)^H = AG$$

(d) $$(GA)^H = GA,$$

where the superscript H denotes the complex conjugate transpose. The matrix G is called

  (i) a g-inverse of A, denoted by A^-, if (a) holds,
 (ii) a *reflexive* g-inverse of A, denoted by A_R^-, if both (a) and (b) hold,
(iii) a *least-squares* g-inverse of A, denoted by A_L^-, if both (a) and (c) hold,
(iv) a *minimum-norm* g-inverse of A, denoted by A_M^-, if both (a) and (d) hold, and
 (v) the *Moore–Penrose* g-inverse of A, denoted by A^+, if (a), (b), (c), and (d) all hold.

2 Properties of g-inverses

In this section we show that, for an arbitrary $m \times n$ matrix A, each type of g-inverse in Definition 1.3 exists, and in addition, we prove that the Moore–Penrose g-inverse of A is unique.

2.1. **Theorem.** *If A^- exists for a square nonsingular matrix A, then $A^- = A^{-1}$.*

PROOF. Let A be a square nonsingular matrix. Also, let A^- exist such that

$$AA^-A = A.$$

If we multiply on the left and on the right by A^{-1} we obtain

$$A^- = A^{-1}. \qquad \square$$

It is well known (see Boullion and Odell [1971], p. 2, for example) that for an arbitrary matrix A, nonsingular (square) matrices P and Q (not necessarily unique) exist such that

$$R = PAQ$$

(2.2)
$$= \begin{bmatrix} I_r & 0 \\ 0 & 0 \end{bmatrix},$$

where r is the rank of A and I_r is the identity matrix of order r. This is called a *diagonal reduction* of A. (We use the symbol 0 for the three zero blocks in R even though the dimensions of the blocks are not necessarily the same). If $r = m$, the two lower zero blocks are absent, and if $r = n$, the two right zero blocks are absent.

If we transpose R and replace the three zero blocks by arbitrary block matrices U, V, and W (of the correct dimensions), we obtain

$$(2.3) \qquad \hat{R} = \begin{bmatrix} I_r & U \\ V & W \end{bmatrix}.$$

2.4. Theorem. *The matrix $G = Q\hat{R}P$ is a g-inverse of A.*

PROOF. From (2.2), $A = P^{-1}RQ^{-1}$ and we can write

$$AGA = (P^{-1}RQ^{-1})(Q\hat{R}P)(P^{-1}RQ^{-1})$$
$$= P^{-1}R\hat{R}RQ^{-1}$$
$$= P^{-1}RQ^{-1}$$
$$= A.$$

Hence, (a) in Definition 1.3 is satisfied. $\qquad\qquad\square$

If we assign the particular values $U = 0$, $V = 0$, and $W = 0$ (where the symbol 0 represents a zero matrix of the appropriate dimensions in each case), then $\hat{R}$ is merely R^T.

2.5. Theorem. *If $G = QR^T P$, then G is a reflexive g-inverse of A.*

PROOF. From Theorem 2.4, $AGA = A$. Likewise,

$$GAG = (QR^T P)(P^{-1}RQ^{-1})(QR^T P)$$
$$= QR^T RR^T P$$
$$= QR^T P$$
$$= G.$$

Hence, both (a) and (b) in Definition 1.3 are satisfied. $\qquad\qquad\square$

2.6. Theorem. *If A is an arbitrary matrix, and if A^- is a g-inverse of A, then*

(i) $(A^-)^H$ *is a g-inverse of A^H.*
(ii) $(1/\lambda)A^-$ *is a g-inverse of λA, if $\lambda \neq 0$.*
(iii) rank $A^- \geq$ rank A.
(iv) AA^- *and A^-A are both idempotent and have the same rank as A.*

PROOF.

$$AA^-A = A$$

implies

$$A^H(A^-)^H A^H = A^H,$$

and so (i) is proved. If $\lambda \neq 0$,

$$(\lambda A)((1/\lambda)A^-)(\lambda A) = \lambda(AA^- A)$$
$$= \lambda A,$$

and so (ii) is proved. Since the rank of the product of matrices is less than or equal to the rank of each factor,

$$AA^- A = A,$$

implies

$$\text{rank } A^- \geq \text{rank } A,$$

and this proves (iii). Obviously

$$(AA^-)^2 = (AA^- A)A^-$$
$$= AA^-,$$

and

$$(A^- A)^2 = A^-(AA^- A)$$
$$= A^- A,$$

and so both AA^- and $A^- A$ are idempotent. Finally,

$$\text{rank } AA^- \leq \text{rank } A$$

and, since $(AA^-)A = A$,

$$\text{rank } A \leq \text{rank } AA^-.$$

Thus,

$$\text{rank } A = \text{rank } AA^-.$$

In a similar manner we can prove that

$$\text{rank } A = \text{rank } A^- A,$$

and this proves (iv). $\square$

Since each type of generalized inverse of A listed in Definition 1.3 satisfies equation (a) in that definition, and since equation (a) is the only hypothesis in Theorem 2.6, we have the obvious corollary.

2.7. Corollary. *Theorem 2.6 is true for each of the g-inverses A_R^-, A_L^-, A_M^-, and A^+.*

2.8. Theorem. *If Y and Z are g-inverses of A, then $G = YAZ$ is a reflexive g-inverse of A.*

PROOF. Since $AYA = A = AZA$, we can write

$$AGA = A(YAZ)A$$
$$= (AYA)ZA$$
$$= AZA$$
$$= A.$$

Also,

$$GAG = (YAZ)A(YAZ)$$
$$= Y(AZA)(YAZ)$$
$$= Y(AYA)Z$$
$$= G.$$ $\quad\square$

2.9. **Theorem.** *For an arbitrary matrix A,*

$$G = (A^H A)^- A^H$$

is a reflexive, least-squares g-inverse of A, and

$$\hat{G} = A^H (A A^H)^-$$

is a reflexive, minimum-norm g-inverse of A.

PROOF. We leave the proof as an exercise for the reader. $\quad\square$

2.10. **Theorem.** *Let A be an arbitrary matrix, let X be a minimum-norm g-inverse of A, and let Y be a least-squares g-inverse of A. Then $G = XAY$ is a Moore–Penrose g-inverse of A.*

PROOF. From Theorem 2.8, G is a reflexive g-inverse of A, that is

$$AGA = A$$

and

$$GAG = G.$$

Thus, we have satisfied (a) and (b) of Definition 1.3.

Since X and Y are minimum-norm and least squares g-inverses, respectively, we have

$$AXA = A$$
$$AYA = A$$
$$(XA)^H = XA$$

and

$$(AY)^H = AY.$$

Thus,

$$AG = A(XAY)$$
$$= AY$$

and

$$GA = (XAY)A$$
$$= XA.$$

Therefore

$$(AG)^H = (AY)^H$$
$$= AY$$
$$= AG$$

and

$$(GA)^H = (XA)^H$$
$$= XA$$
$$= GA.$$

Hence, (c) and (d) of Definition 1.3 are satisfied and we have proved that

$$G = A^+.$$ □

With these theorems we have shown the existence of each type of g-inverse in Definition 1.3. We next prove the uniqueness of the Moore–Penrose g-inverse.

2.11. Theorem. *The Moore–Penrose g-inverse of A is unique.*

PROOF. Let G and $\underline{G}$ be two Moore–Penrose g-inverses of A. Then

$$
\begin{array}{ccc}
AGA = A & & A\underline{G}A = A \\
GAG = G & & \underline{G}A\underline{G} = \underline{G} \\
(AG)^H = AG & \text{and} & (A\underline{G})^H = A\underline{G} \\
(GA)^H = GA & & (\underline{G}A)^H = \underline{G}A.
\end{array}
$$

Consequently,

$$G = GAG$$
$$= G(AG)^H$$
$$= GG^H A^H.$$

Since

$$A^H = A^H \underline{G}^H A^H,$$

we have

$$G = GG^H(A^H\underline{G}^H A^H)$$
$$= G(AG)^H(A\underline{G})^H$$
$$= GAGA\underline{G}$$
$$= GA\underline{G}.$$

In a similar manner we can show that

$$\underline{G} = GA\underline{G}.$$

Thus,

$$\underline{G} = G. \qquad \square$$

2.12. **Theorem.** *If A belongs to certain special classes of matrices, the Moore–Penrose g-inverse can be computed as follows.*

(i) *If A is square and nonsingular, $A^+ = A^{-1}$.*

(ii) *If $A = 0$, $A^+ = 0$, where the two zero matrices are the transpose of each other.*

(iii) *If A is an $m \times n$ matrix of rank n,*

$$A^+ = (A^H A)^{-1} A^H.$$

(iv) *If A is an $m \times n$ matrix of rank m,*

$$A^+ = A^H(AA^H)^{-1}.$$

PROOF. The proof is left as an exercise for the reader. $\qquad \square$

2.13. **Theorem.** *If A is an $m \times n$ matrix of rank r, there exist matrices B and C such that $A = BC$, where B is an $m \times r$ matrix of rank r and C is an $r \times n$ matrix of rank r. (Such a factorization is called a full-rank factorization of A.) Then,*

$$A^+ = C^H(B^H A C^H)^{-1} B^H.$$

PROOF. See Ben-Israel and Greville [1974], p. 23. $\qquad \square$

2.14. **Theorem.** *If A is an arbitrary matrix, then*

(i) $$(A^+)^+ = A$$

(ii) $$(A^H)^+ = (A^+)^H$$

(iii) $$(A^T)^+ = (A^+)^T$$

(iv) $$A^+ = (A^H A)^+ A^H$$

(v) $$A^+ = A^H(AA^H)^+$$

PROOF. The proof is left as an exercise for the reader. $\qquad\square$

EXERCISES III.2

1. Prove Theorem 2.9.

2. Prove Theorem 2.12.

3. Prove Theorem 2.14.

4. Prove that if A is Hermitian and idempotent, then $A^+ = A$.

5. Prove that if U and V are unitary, then

$$(UAV)^+ = V^H A^+ U^H$$

for any matrix A for which UAV is defined.

6. Let s be a scalar and define

$$s^+ = \begin{cases} s^{-1} & \text{for } s \neq 0 \\ 0 & \text{for } s = 0. \end{cases}$$

Prove that if $D = \operatorname{diag}(d_1, d_2, \ldots, d_n)$, then $D^+ = \operatorname{diag}(d_1^+, d_2^+, \ldots, d_n^+)$.

3 Applications of g-inverses

Consider the system of m linear algebraic equations in n unknowns

(3.1)
$$\begin{cases} a_{11}x_1 + a_{12}x_2 + \cdots + a_{1n}x_n = b_1 \\ a_{21}x_1 + a_{22}x_2 + \cdots + a_{2n}x_n = b_2 \\ \cdots\cdots\cdots\cdots\cdots\cdots\cdots\cdots\cdots\cdots \\ a_{m1}x_1 + a_{m2}x_2 + \cdots + a_{mn}x_n = b_m \end{cases}$$

which we write in the form

(3.2) $$Ax = b.$$

Here A is an $m \times n$ matrix, x is an n-vector, and b is an m-vector. We shall discuss this system of equations in terms of the various g-inverses of A. If (3.2) has a solution, (3.2) is called *consistent*. We point out that all of the definitions and theorems of this section are in Rao and Mitra [1971], Chapters 2 and 3, and so the proofs of the theorems are omitted.

3.3. **Theorem.** *Let A^- be any g-inverse of the coefficient matrix A in (3.2). Then (3.2) is consistent if and only if $AA^-b = b$, in which case the most general solution is*

$$x = A^-b + (I - A^-A)z,$$

where z is an arbitrary n-vector.

3.4. Definition. Let $Ax = b$ be a consistent system of linear algebraic equations. A solution y is said to be a *minimum-norm solution* if $\|y\|_2$ is minimum among the norms of all solutions, that is, if

$$\|y\|_2 \leq \|x\|_2$$

for every solution x.

3.5. Theorem. *Let the system $Ax = b$ be consistent, and let A_M^- be a minimum-norm g-inverse of A. Then $y = A_M^- b$ is a minimum-norm solution of $Ax = b$.*

3.6. Theorem. *The minimum-norm solution is unique (although A_M^- may not be unique).*

3.7. Definition. Let $Ax = b$ be an inconsistent system of linear algebraic equations. A vector y is said to be a *least-squares solution* of the system if

$$\|Ay - b\|_2 \leq \|Ax - b\|_2$$

for all x.

3.8. Theorem. *Let $Ax = b$ be an inconsistent system of linear algebraic equations and let A_L^- be a least-squares g-inverse of A. Then $y = A_L^- b$ is a least-squares solution of $Ax = b$.*

Theorems 3.5 and 3.6 provide us with the minimum-norm solution if $Ax = b$ is consistent, and Theorem 3.8 provides us with a least-squares solution if $Ax = b$ is inconsistent. The least-squares solution is not unique, however, and we wish to choose, from among the least-squares solutions, one with a minimum norm.

3.9. Definition. Let $Ax = b$ be an inconsistent system of linear algebraic equations. Let y be a least-squares solution of the system such that

$$\|y\|_2 \leq \|x\|_2$$

for every least squares solution x. Then y is said to be a *minimum-norm least-squares solution* of $Ax = b$.

3.10. Theorem. *Let $Ax = b$ be an inconsistent system of linear algebraic equations and let A^+ be the Moore–Penrose g-inverse of A. Then $y = A^+ b$ is a minimum-norm least-squares solution of $Ax = b$.*

3.11. Theorem. *The minimum-norm least-squares solution of $Ax = b$ is unique.*

4 Exact Computation of A^+ if A Is a Rational Matrix

We choose the computation of the Moore–Penrose g-inverse as an application of error-free computation because it is the most useful of the g-inverses (for example, it is unique) and because it is rather difficult to compute using ordinary computer arithmetic. It should be pointed out that, for certain special classes of matrices, Theorem 2.12 can be useful in computing A^+.

For an arbitrary matrix A, many of the algorithms require that we be able to recognize the rank of A numerically, and this is an extremely difficult task in the presence of rounding errors because we must determine whether or not a small quantity in our computer represents zero.

If A is a (square) nonsingular matrix, the inverse is a continuous function of the matrix elements, that is,

$$(4.1) \qquad \lim_{E \to 0} (A + E)^{-1} = A^{-1}.$$

On the other hand, if A is arbitrary, A^+ is not necessarily a continuous function of the matrix elements. See Stewart [1969], for example.

4.2. EXAMPLE. See Rao [1975]. Let A be the matrix

$$A = \begin{bmatrix} 1 & 1 \\ 1 & 1 + \varepsilon \end{bmatrix}.$$

Then $A^+ = \frac{1}{4}A$ for $\varepsilon = 0$, but when $\varepsilon \neq 0$,

$$A^+ = \begin{bmatrix} 1 + 1/\varepsilon & -1/\varepsilon \\ -1/\varepsilon & 1/\varepsilon \end{bmatrix}$$

$$= A^{-1}.$$

The existence of such discontinuities presents further difficulties in computation. It is obvious, therefore, that it is highly desirable to compute A^+ using error-free computation such as residue or finite-segment p-adic arithmetic.

In this chapter we use residue arithmetic and apply it to the computation of A^+ using an algorithm based on the diagonal reduction of the square matrix $M = (AA^T)^2$ to the form (2.2). Rao, Subramanian and Krishnamurthy [1976] refer to (2.2) as the Hermite canonical form* and they refer to their algorithm as the Hermite algorithm. It uses the equation

$$(4.3) \qquad A^+ = A^T M_R^- AA^T,$$

where the computation of M_R^- is based on (2.2) and Theorem 2.5.

We assume that the matrix A has only integer elements since rational

---

* For a different definition of the Hermite canonical form see Ben-Israel and Greville [1974], or Rao and Mitra [1971].

elements can be converted to integers by proper scaling. (See Theorem 2.6, part (ii) and Corollary 2.7, for example.)

Following a discussion of the Hermite algorithm, we discuss two additional algorithms; an algorithm due to Greville [1960], and an algorithm due to Decell [1965] based on Leverrier's method. (See Faddeev and Faddeeva [1963].)

The Hermite algorithm

Given an arbitrary matrix A, we can easily form the (square) matrix

$$(4.4) \qquad M = (AA^T)^2.$$

The main thrust of the Hermite algorithm involves the computation of M_R^- using (2.2) and Theorem 2.5. Once we have M_R^- we compute A^+ using the matrix multiplications indicated in (4.3).

4.5. Definition. A matrix is said to be in *row-echelon form* if it satisfies the following conditions:

(i) Each zero row follows all nonzero rows.
(ii) The leading nonzero element in each nonzero row is 1.
(iii) Any column containing such a leading element has zeros elsewhere in the column.
(iv) Consider the nonzero rows. If $i < j$, the leading nonzero element in row i appears to the left of the leading nonzero element in row j.

4.6. Definition. A matrix which possesses all of the above properties except (iii) is said to be in *simple row-echelon form*.

We can define *column-echelon form* and *simple column-echelon form* in an analogous manner.

4.7. EXAMPLE. Let

$$A = \begin{bmatrix} 1 & 0 & 1 & 2 & 0 \\ 0 & 0 & 1 & 3 & 2 \\ 0 & 0 & 0 & 0 & 0 \end{bmatrix}.$$

This matrix is in simple row-echelon form. However, if rows one and two are interchanged, the resulting matrix is not in simple row-echelon form because (iv) no longer holds.

We now return to the matrix $M = (AA^T)^2$. Clearly, there exists a nonsingular matrix E such that

$$(4.8) \qquad EM = M_1$$

where M_1 is in simple row-echelon form. Obviously, M_1^T is in simple column-echelon form. Similarly, there exists a nonsingular matrix F such that

(4.9)
$$FM_1^T = R$$
$$= \begin{bmatrix} I_k & 0 \\ 0 & 0 \end{bmatrix},$$

where k is the rank of M, and the square matrix R is in both row-echelon and column-echelon form. Thus, we have

(4.10)
$$FM^T E^T = R,$$

which implies (since R is symmetric)

(4.11)
$$R = EMF^T.$$

This equation is in the form (2.2) with $E = P$ and $F^T = Q$. Thus, using Theorem 2.5, we have

(4.12)
$$M_R^- = F^T R E.$$

4.13. **Remark.** If we write the product on the right in partitioned form, we observe that

$$M_R^- = \begin{bmatrix} A & B \\ C & D \end{bmatrix} \begin{bmatrix} I_r & 0 \\ 0 & 0 \end{bmatrix} \begin{bmatrix} G & H \\ K & L \end{bmatrix}$$
$$= \begin{bmatrix} A & 0 \\ C & 0 \end{bmatrix} \begin{bmatrix} G & H \\ 0 & 0 \end{bmatrix}$$

which implies only the first r columns of F^T and only the first r rows of E are needed in forming the reflexive g-inverse of M.

In order to compute the matrices E and F we proceed as follows. Successive premultiplication of the augmented matrix $[M : I]$ by elementary row transformations produces the augmented matrix $[M_1 : E]$, that is,

(4.14)
$$E_s \cdots E_2 E_1 [M : I] = [M_1 : E],$$

Using a similar procedure on M_1^T, we obtain

(4.15)
$$F_t \cdots F_2 F_1 [M_1^T : I] = [R : F]$$

where F_i represents an elementary row transformation for $i = 1, 2, \ldots, t$.

Single-modulus residue arithmetic

Since we are assuming that A has integer elements, it is clear that $M = (AA^T)^2$ also has integer elements. However, M_R^- and A^+ will not have integer elements, in general. If we let

$$(4.16) \qquad A^+ = (\alpha_{ij}),$$

then the matrix whose elements are the integers

$$(4.17) \qquad a_{ij} = |\alpha_{ij}|_p$$

is denoted by

$$(4.18) \qquad |A^+|_p = (a_{ij}).$$

4.19. EXAMPLE. See Rao, Subramanian, and Krishnamurthy [1976]. Let A be the singular matrix

$$A = \begin{bmatrix} 1 & 0 & 1 \\ 1 & 1 & 0 \\ 1 & 0 & 1 \end{bmatrix}.$$

Then $M = (AA^T)^2$ is the matrix

$$M = \begin{bmatrix} 9 & 6 & 9 \\ 6 & 6 & 6 \\ 9 & 6 & 9 \end{bmatrix}.$$

Before we compute A^+ using residue arithmetic, we demonstrate the computation using ordinary rational arithmetic. We can follow a procedure similar to that used in Example 7.5, Chapter II, and obtain

$$\begin{bmatrix} 9 & 6 & 9 & 1 & 0 & 0 \\ 6 & 6 & 6 & 0 & 1 & 0 \\ 9 & 6 & 9 & 0 & 0 & 1 \end{bmatrix} \to \begin{bmatrix} 1 & \frac{2}{3} & 1 & \frac{1}{9} & 0 & 0 \\ 6 & 6 & 6 & 0 & 1 & 0 \\ 9 & 6 & 9 & 0 & 0 & 1 \end{bmatrix}$$

$$\to \begin{bmatrix} 1 & \frac{2}{3} & 1 & \frac{1}{9} & 0 & 0 \\ 0 & 2 & 0 & -\frac{2}{3} & 1 & 0 \\ 0 & 0 & 0 & -1 & 0 & 1 \end{bmatrix} \to \begin{bmatrix} 1 & \frac{2}{3} & 1 & \frac{1}{9} & 0 & 0 \\ 0 & 1 & 0 & -\frac{1}{3} & \frac{1}{2} & 0 \\ 0 & 0 & 0 & -1 & 0 & 1 \end{bmatrix},$$

which implies

$$M_1 = \begin{bmatrix} 1 & \frac{2}{3} & 1 \\ 0 & 1 & 0 \\ 0 & 0 & 0 \end{bmatrix},$$

and

$$E = \begin{bmatrix} \frac{1}{9} & 0 & 0 \\ -\frac{1}{3} & \frac{1}{2} & 0 \\ -1 & 0 & 1 \end{bmatrix}.$$

Next, we augment M_1^T and obtain

$$
\begin{bmatrix} 1 & 0 & 0 & 1 & 0 & 0 \\ \frac{2}{3} & 1 & 0 & 0 & 1 & 0 \\ 1 & 0 & 0 & 0 & 0 & 1 \end{bmatrix} \rightarrow \begin{bmatrix} 1 & 0 & 0 & 1 & 0 & 0 \\ 0 & 1 & 0 & -\frac{2}{3} & 1 & 0 \\ 0 & 0 & 0 & -1 & 0 & 1 \end{bmatrix},
$$

which implies

$$
R = \begin{bmatrix} 1 & 0 & 0 \\ 0 & 1 & 0 \\ 0 & 0 & 0 \end{bmatrix},
$$

and

$$
F = \begin{bmatrix} 1 & 0 & 0 \\ -\frac{2}{3} & 1 & 0 \\ -1 & 0 & 1 \end{bmatrix}.
$$

Consequently, if we use the result in Remark 4.13, we need only the first two columns of F^T and the first two rows of E to form

$$
M_R^- = \begin{bmatrix} 1 & -\frac{2}{3} \\ 0 & 1 \\ 0 & 0 \end{bmatrix} \begin{bmatrix} \frac{1}{9} & 0 & 0 \\ -\frac{1}{3} & \frac{1}{2} & 0 \end{bmatrix}
$$

$$
= \begin{bmatrix} \frac{1}{3} & -\frac{1}{3} & 0 \\ -\frac{1}{3} & \frac{1}{2} & 0 \\ 0 & 0 & 0 \end{bmatrix}.
$$

From (4.3), we obtain

$$
A^+ = A^T M_R^- A A^T
$$

$$
= \frac{1}{6} \begin{bmatrix} 1 & 2 & 1 \\ -1 & 4 & -1 \\ 2 & -2 & 2 \end{bmatrix}.
$$

Now we compute M_R^-, and then A^+, using single-modulus residue arithmetic. Rao et al. [1976] propose that the prime modulus p satisfy

$$
p > 2 \prod_{j=1}^{m} \|c_j\|_2
$$

where $\|c_j\|_2$ is the Euclidean norm of the jth column of M. (If M contains a zero column, that column is not used in forming the product.) Using this criterion we can safely choose $p = 4357$.

As before, we transform the augmented matrix $[\,|M\,|_p : I\,]$ into $[\,|M_1\,|_p : |E\,|_p\,]$ as follows, where the arithmetic is carried out modulo p.

$$\begin{bmatrix} 9 & 6 & 9 & 1 & 0 & 0 \\ 6 & 6 & 6 & 0 & 1 & 0 \\ 9 & 6 & 9 & 0 & 0 & 1 \end{bmatrix} \rightarrow \begin{bmatrix} 1 & 1453 & 1 & 3873 & 0 & 0 \\ 6 & 6 & 6 & 0 & 1 & 0 \\ 9 & 6 & 9 & 0 & 0 & 1 \end{bmatrix}$$

$$\rightarrow \begin{bmatrix} 1 & 1453 & 1 & 3873 & 0 & 0 \\ 0 & 2 & 0 & 2904 & 1 & 0 \\ 0 & 0 & 0 & 4356 & 0 & 1 \end{bmatrix} \rightarrow \begin{bmatrix} 1 & 1453 & 1 & 3873 & 0 & 0 \\ 0 & 1 & 0 & 1452 & 2179 & 0 \\ 0 & 0 & 0 & 4356 & 0 & 1 \end{bmatrix}$$

which implies

$$|M_1|_{4357} = \begin{bmatrix} 1 & 1453 & 1 \\ 0 & 1 & 0 \\ 0 & 0 & 0 \end{bmatrix}$$

and

$$|E|_{4357} = \begin{bmatrix} 3873 & 0 & 0 \\ 1452 & 2179 & 0 \\ 4356 & 0 & 1 \end{bmatrix}.$$

We now transform the augmented matrix $[|M_1^T|_p : I]$ into $[|R|_p : |F|_p]$. The computation follows.

$$\begin{bmatrix} 1 & 0 & 0 & 1 & 0 & 0 \\ 1453 & 1 & 0 & 0 & 1 & 0 \\ 1 & 0 & 0 & 0 & 0 & 1 \end{bmatrix} \rightarrow \begin{bmatrix} 1 & 0 & 0 & 1 & 0 & 0 \\ 0 & 1 & 0 & 2904 & 1 & 0 \\ 0 & 0 & 0 & 4356 & 0 & 1 \end{bmatrix}$$

which implies

$$|R|_{4357} = \begin{bmatrix} 1 & 0 & 0 \\ 0 & 1 & 0 \\ 0 & 0 & 0 \end{bmatrix}$$

and

$$|F|_{4357} = \begin{bmatrix} 1 & 0 & 0 \\ 2904 & 1 & 0 \\ 4356 & 0 & 1 \end{bmatrix}.$$

At this point, we use the result in Remark 4.13 and obtain

$$|M_R^-|_{4357} = \begin{bmatrix} 1 & 2904 \\ 0 & 1 \\ 0 & 0 \end{bmatrix} \begin{bmatrix} 3873 & 0 & 0 \\ 1452 & 2179 & 0 \end{bmatrix}$$

$$= \begin{bmatrix} 2905 & 1452 & 0 \\ 1452 & 2179 & 0 \\ 0 & 0 & 0 \end{bmatrix}.$$

From (4.3), we have

$$|A^+|_p = |A^T|_p |M_R^-|_p |AA^T|_p.$$

Hence,

$$|A^+|_{4357} = \begin{bmatrix} 3631 & 2905 & 3631 \\ 726 & 1453 & 726 \\ 2905 & 1452 & 2905 \end{bmatrix}.$$

Finally, we use Algorithm 6.26, Chapter I, to map the integers 3631, 2905, 726, 1452, and 1453 onto their rational equivalents.

	4357	0			4357	0
	3631	1			2905	1
1	726	−1		1	1452	−1
5	1	6		2	1	3
726	0	−4357		1452	0	−4357

	4357	0			4357	0
	726	1			1452	1
6	1	−6		3	1	−3
726	0	4357		1452	0	4357

	4357	0
	1453	1
2	1451	−2
1	2	3
725	1	−2177
2	0	4357

Hence,

$$A^+ = \begin{bmatrix} \frac{1}{6} & \frac{1}{3} & \frac{1}{6} \\ -\frac{1}{6} & \frac{2}{3} & -\frac{1}{6} \\ \frac{1}{3} & -\frac{1}{3} & \frac{1}{3} \end{bmatrix}.$$

Multiple-modulus residue arithmetic

In Example 4.19 we used the single modulus $p = 4357$ for a matrix of order three. Obviously, as the order of the matrix increases the value of p will have to increase. We can use multiple-modulus residue arithmetic in order to keep our intermediate results small, as the following example demonstrates.

4.20. EXAMPLE. See Rao [1975]. Let A be the singular matrix

$$A = \begin{bmatrix} 1 & 1 \\ 2 & 2 \end{bmatrix}.$$

Then

$$M = \begin{bmatrix} 20 & 40 \\ 40 & 80 \end{bmatrix}.$$

Using the criterion

$$p > 2 \prod_{j=1}^{2} \|c_j\|_2$$

we find that p must be at least 8000 if we use a single modulus. We choose to use two moduli whose product exceeds 8000, namely $p_1 = 101$ and $p_2 = 103$, and in each case $|A|_{p_i} = A$ and $|M|_{p_i} = M$.

The case $p_1 = 101$

The computation leading to $|M_1|_{101}$ and $|E|_{101}$ follows.

$$\begin{bmatrix} 20 & 40 & 1 & 0 \\ 40 & 80 & 0 & 1 \end{bmatrix} \rightarrow \begin{bmatrix} 1 & 2 & 96 & 0 \\ 40 & 80 & 0 & 1 \end{bmatrix} \rightarrow \begin{bmatrix} 1 & 2 & 96 & 0 \\ 0 & 0 & 99 & 1 \end{bmatrix}$$

which implies

$$|M_1|_{101} = \begin{bmatrix} 1 & 2 \\ 0 & 0 \end{bmatrix}$$

and

$$|E|_{101} = \begin{bmatrix} 96 & 0 \\ 99 & 1 \end{bmatrix}.$$

We now compute $|R|_{101}$ and $|F|_{101}$ as follows.

$$\begin{bmatrix} 1 & 0 & 1 & 0 \\ 2 & 0 & 0 & 1 \end{bmatrix} \rightarrow \begin{bmatrix} 1 & 0 & 1 & 0 \\ 0 & 0 & 99 & 1 \end{bmatrix}$$

which implies

$$|R|_{101} = \begin{bmatrix} 1 & 0 \\ 0 & 0 \end{bmatrix}$$

and

$$|F|_{101} = \begin{bmatrix} 1 & 0 \\ 99 & 1 \end{bmatrix}.$$

At this point, we use the result in Remark 4.13 and obtain

$$|M_R^-|_{101} = \begin{bmatrix} 1 \\ 0 \end{bmatrix} \begin{bmatrix} 96 & 0 \end{bmatrix}$$

$$= \begin{bmatrix} 96 & 0 \\ 0 & 0 \end{bmatrix}.$$

From (4.3), we have

$$|A^+|_{101} = |A^T|_{101} |M_R^-|_{101} |AA^T|_{101}.$$

Hence,

$$|A^+|_{101} = \begin{bmatrix} 91 & 81 \\ 91 & 81 \end{bmatrix}.$$

The case $p_2 = 103$

By repeating the steps above, for the case $p_2 = 103$, we find that

$$|A^+|_{103} = \begin{bmatrix} 31 & 62 \\ 31 & 62 \end{bmatrix}.$$

Combining the results

We now use the notation of Section 3, Chapter I, and write the base vector

$$\beta = [101, 103].$$

Hence,

$$|A|_\beta^+ = \begin{bmatrix} [91, 31] & [81, 62] \\ [91, 31] & [81, 62] \end{bmatrix}.$$

Since multiple-modulus residue arithmetic is equivalent to single-modulus residue arithmetic with the single modulus (not a prime) $p = p_1 p_2$, and since, in this example, $p = 10403$ we now find $|A^+|_{10403}$ using the mixed-radix procedure of Section 4, Chapter I.

First we convert the representation

$$|\alpha_{11}|_\beta = [91, 31]$$

to the integer $|\alpha_{11}|_{10403}$, which has the unique mixed-radix representation

$$|\alpha_{11}|_{10403} = x_0 + x_1(101)$$

where

$$0 \le x_0 < 101,$$

and

$$0 \leq x_1 < 103.$$

From (4.15), Chapter I, the first mixed-radix digit x_0 is the same as the first residue digit. Hence

$$x_0 = 91.$$

Using the method of Problem 4.39, Chapter I, we obtain

$$x_1 = 30,$$

as the second mixed-radix digit. Thus,

$$|\alpha_{11}|_{10403} = 3121.$$

In a similar manner we find that

$$|\alpha_{12}|_{10403} = 6242.$$

Thus, since $|\alpha_{11}|_{10403} = |\alpha_{21}|_{10403}$ and $|\alpha_{12}|_{10403} = |\alpha_{22}|_{10403}$,

$$|A^+|_{10403} = \begin{bmatrix} 3121 & 6242 \\ 3121 & 6242 \end{bmatrix}.$$

We now use Algorithm 6.26, Chapter I, to map the integers 3121 and 6242 onto their rational equivalents.

	10403	0			10403	0
	3121	1			6242	1
3	1040	−3	1		4161	−1
3	1	10	1		2081	2
1040	0	−10403	1		2080	−3
			1		1	5
			2080		0	−10403

Hence,

$$A^+ = \begin{bmatrix} \frac{1}{10} & \frac{1}{5} \\ \frac{1}{10} & \frac{1}{5} \end{bmatrix}.$$

Greville's algorithm

This algorithm (see Greville [1960]) is described in Krishnamurthy and Sen [1976], p. 191. A recursive procedure is used in which the $m \times n$ matrix

$$(4.21) \qquad A = [a_1, a_2, \ldots, a_n]$$

is created by beginning with the first column, a_1, and appending a new column at each step. Thus,

(4.22) $$A_i = [a_1, a_2, \ldots, a_i]$$

can also be written in the form

(4.23) $$A_i = [A_{i-1} : a_i] \qquad i = 2, 3, \ldots, n.$$

The key theorem (see Rao, Subramanian, and Krishnamurthy [1976], p. 161 and Rao and Mitra [1971], p. 64) relates A_i^+ to A_{i-1}^+.

4.24. Theorem. *Let A_{i-1} be an $m \times (i-1)$ matrix and let a_i be an m-vector. Then*

$$A_i^+ = \begin{bmatrix} A_{i-1}^+ - d_i b_i^T \\ b_i^T \end{bmatrix}$$

where

$$d_i = A_{i-1}^+ a_i,$$

and

$$b_i = \begin{cases} \dfrac{1}{c_i^T c_i} c_i & \text{if } c_i \neq 0 \\[2ex] \dfrac{1}{1 + d_i^T d_i} (A_{i-1}^+)^T d_i & \text{if } c_i = 0 \end{cases}$$

where

$$c_i = a_i - A_{i-1} d_i.$$

PROOF. See Greville [1960].   □

To begin the recursion, we use

(4.25) $$A_1^+ = \begin{cases} a_1^T & \text{if } a_1 = 0 \\[2ex] \dfrac{1}{a_1^T a_1} a_1^T & \text{if } a_1 \neq 0 \end{cases}$$

The value of c_i is critical in computing b_i. If ordinary floating-point arithmetic is used, it is extremely difficult (in the presence of rounding errors) to determine whether or not each component of c_i is exactly zero. Hence, the algorithm suffers from numerical instability. Obviously, error-free computation can be quite useful in overcoming this difficulty.

The Decell–Leverrier algorithm

Stallings and Boullion [1972] employ residue arithmetic in computing A^+ using an algorithm due to Decell [1965] which is based on Leverrier's method. See Faddeev and Faddeeva [1963]. The following theorem is the basis for this algorithm.

4.26. Theorem. *Let A be any $m \times n$ complex matrix and let $B = AA^H$ have the characteristic polynomial*

$$B(\lambda) = (-1)^m(a_0\lambda^m + a_1\lambda^{m-1} + \ldots + a_m),$$

with $a_0 = 1$. If k is the largest integer such that $a_k \neq 0$, then

$$A^+ = \begin{cases} -\dfrac{1}{a_k}A^H(a_0 B^{k-1} + a_1 B^{k-2} + \ldots + a_{k-1}I) & k > 0 \\[2mm] 0 & k = 0. \end{cases}$$

Also, rank $A = k$.

PROOF. See Decell [1965].   □

Based on this theorem, Stallings and Boullion describe an *exact* method for computing rank A and A^+ in the special case where A is real with integer elements.

Step 1. Compute $B = AA^T$.
Step 2. Find the characteristic polynomial $B(\lambda)$. This is done by computing the coefficients $a_0, a_1, \ldots, a_m$ by using Leverrier's method as shown below.

If $\lambda_1, \lambda_2, \ldots, \lambda_m$ are the eigenvalues of B, and if we define

$$(4.27) \qquad\qquad s_k = \sum_{i=1}^m \lambda_i^k$$

for $1 \leq k \leq m$, then

$$(4.28) \qquad\qquad s_k = \mathrm{tr}(B^k).$$

If we use Newton's formulas (see Faddeev and Faddeeva [1963], p. 260) we can write

$$(4.29a) \quad \begin{bmatrix} 1 & & & & \\ s_1 & 2 & & & \\ s_2 & s_1 & 3 & & \\ s_3 & s_2 & s_1 & 4 & \\ \multicolumn{5}{c}{\dotfill} \\ s_{m-1} & s_{m-2} & s_{m-3} & s_{m-4} & \cdots & m \end{bmatrix} \begin{bmatrix} a_1 \\ a_2 \\ a_3 \\ a_4 \\ \vdots \\ a_m \end{bmatrix} = - \begin{bmatrix} s_1 \\ s_2 \\ s_3 \\ s_4 \\ \vdots \\ s_m \end{bmatrix}$$

or, simply,

$$(4.29b) \qquad\qquad La = s.$$

The computation of a_k from (4.29) and the computation of A^+ proceed as follows:

(i) We compute s_k, for $1 \leq k \leq m$, using (4.28). This requires the computation of the powers of B and the traces of the powers of B.

(ii) Using these values of s_k we solve (4.29) using "forward substitution," since L is a lower triangular matrix.*

Step 3. Determine A^+ using Theorem 4.26.

Stallings and Boullion, however, use a sequential algorithm in computing a_k which involves the construction of a sequence of matrices $A_0, A_1, \ldots, A_k$ as follows:

$$
\begin{aligned}
A_0 &= 0   & q_0 &= -1   & B_0 &= I \\
A_1 &= AA^T   & q_1 &= \operatorname{tr} A_1 & B_1 &= A_1 - q_1 I \\
A_2 &= A_1 B_1   & q_2 &= \tfrac{1}{2} \operatorname{tr} A_2 & B_2 &= A_2 - q_2 I \\
A_3 &= A_1 B_2   & q_3 &= \tfrac{1}{3} \operatorname{tr} A_3 & B_3 &= A_3 - q_3 I \\
&\;\;\vdots   &   &\;\;\vdots   &   &\;\;\vdots \\
A_k &= A_1 B_{k-1}   & q_k &= \tfrac{1}{k} \operatorname{tr} A_k & B_k &= A_k - q_k I
\end{aligned}
$$

Notice that $q_i = -a_i$ in the expression for A^+ in Theorem 4.26. Consequently, for $k > 0$,

$$
(4.30) \qquad\qquad A^+ = \frac{1}{q_k} A^T B_{k-1}.
$$

Since k is not known *a priori*, the iteration is continued until the matrix product

$$
(4.31) \qquad\qquad A_1 B_k = 0.
$$

Also, since B_{k-1} and A^T turn out to be matrices with integer elements, it is not necessary to store a common denominator in this method. (The constant q_k is the common denominator.)

The critical step in Decell's algorithm involves the determination of whether or not $A_1 B_k = 0$. Obviously, error-free computation is useful in overcoming the numerical instability in this algorithm.

4.32. Remark. We have not illustrated the use of error-free arithmetic by solving numerical examples using the last two algorithms. However, Stallings and Boullion [1972] and Rao, Subramanian, and Krishnamurthy [1976] contain such examples. In addition, Rao et al. demonstrate the superiority of the Hermite algorithm over the other two.

EXERCISE III.4

1. Compute $M = (AA^T)^2$ for

(a)
$$
A = \begin{bmatrix} 1 & 2 & 4 \\ 1 & 0 & 1 \end{bmatrix}
$$

---

*The triangular system (4.29) can be solved by computing L^{-1} using a parallel method due to Krishnamurthy [1983]. (Forward substitution is sequential.) Then $a = L^{-1} s$.

(b)
$$A = \begin{bmatrix} 1 & 0 & 1 & 1 \\ 0 & 1 & -1 & 0 \\ 1 & 1 & 0 & 1 \end{bmatrix}$$

(c)
$$A = \begin{bmatrix} 1 & 1 \\ 2 & 1 \end{bmatrix}$$

2. Compute M_R^- for each matrix in Problem 1.

3. Compute A^+ for each matrix in Problem 1 using ordinary rational arithmetic with the Hermite algorithm.

4. Compute A^+ for each matrix in Problem 1 using residue arithmetic with the Hermite algorithm.

5. Compute A^+ for the matrix in Problem 1(a) using ordinary rational arithmetic with Greville's algorithm.

6. Compute A^+ for the matrix in Problem 1(c) using ordinary rational arithmetic with the Decell–Leverrier algorithm.

5 Failures of Residue Arithmetic and Precautionary Measures

It is necessary to indicate some of the difficulties which are associated with the use of residue arithmetic and to indicate how to take precautionary measures to avoid them.

For an arbitrary (square) matrix over the complex field, the following statements are true.

(a) rank A = rank AA^H = rank $A^H A$.
(b) rank A = rank A^2.
(c) If $AA^H = 0$, then $A = 0$, and conversely.
(d) If $A^H A = 0$, then $A = 0$, and conversely.

If the matrix A is over the real field then A^H becomes A^T in the equations above.

Unfortunately, we cannot guarantee any of the above statements if the matrix A is over a finite field, as may be seen from the following examples.

5.1. EXAMPLE. Consider the matrix

$$A = \begin{bmatrix} 1 & 1 & 1 & 1 & 1 \\ 2 & 2 & 2 & 2 & 2 \\ 3 & 3 & 3 & 3 & 3 \\ 4 & 4 & 4 & 4 & 4 \\ 1 & 1 & 1 & 1 & 1 \end{bmatrix}.$$

where the elements are taken from the finite field $(\mathbb{I}_5, +, \cdot)$. It is easy to verify that

$$\text{rank } A = 1,$$

$$AA^T = 0,$$

$$\text{rank } AA^T = 0,$$

$$A^TA = \begin{bmatrix} 1 & 1 & 1 & 1 & 1 \\ 1 & 1 & 1 & 1 & 1 \\ 1 & 1 & 1 & 1 & 1 \\ 1 & 1 & 1 & 1 & 1 \\ 1 & 1 & 1 & 1 & 1 \end{bmatrix}$$

and

$$\text{rank } A^TA = 1.$$

5.2. EXAMPLE. Consider the matrix

$$A = \begin{bmatrix} 1 & 1 & 1 \\ 1 & 1 & 1 \\ 1 & 1 & 1 \end{bmatrix},$$

where the elements are taken from the finite field $(\mathbb{I}_3, +, \cdot)$. It is easily seen that

$$\text{rank } A = 1,$$

$$AA^T = A^TA = A^2 = 0,$$

and

$$\text{rank } A^2 = 0.$$

From these examples we observe that the rank of AA^T over the real field and the rank of $|AA^T|_p$ may be different, with the possibility that

$$(5.3) \qquad \text{rank } |AA^T|_p \leq \text{rank } AA^T.$$

Since all the formulas for computing A_L^-, A_M^-, and A^+ are based on the computation of AA^T, A^TA, or $(AA^T)^2$, it is clear that there may not be a unique correspondence between $|A_L^-|_p$, $|A_M^-|_p$, and $|A^+|_p$ and their counterparts over the real field. In fact, while using multiple-modulus residue arithmetic for the computation of $|A_L^-|_{p_i}$, $|A_M^-|_{p_i}$, and $|A^+|_{p_i}$, for various primes p_i, it is necessary (in order to obtain A_L^-, A_M^-, and A^+ using the Chinese Remainder Theorem) that the residue forms exist and have the same rank. Consequently, it is necessary that we guarantee that rank $|AA^T|_{p_i}$ be the same for each prime p_i.

One way to guarantee that the ranks are the same is to do the following.

In the computation of A^+ we use (4.12) to obtain M_R^-. Thus, we compute

(5.4) $$|M_R^-|_{p_i} = |F^T RE|_{p_i}$$

and

(5.5) $$|M_R^-|_{p_j} = |F^T RE|_{p_j}$$

for $p_i \neq p_j$ and compare their ranks. If

(5.6) $$\text{rank}\,|M_R^-|_{p_i} < \text{rank}\,|M_R^-|_{p_j},$$

then p_i should be replaced by another modulus. Also, there should be a check to see whether or not a new choice of p_i leads to an increase in rank exceeding the rank obtained for p_j. If this should be the case, then all previously chosen primes should be rejected and the process should be begun anew. To minimize the possibility of this kind of difficulty we should choose only very large prime moduli.

While using the Decell–Leverrier algorithm the following check can be made. Make certain that

$$|a_k|_{p_i} = |a_k|_{p_j} \neq 0$$

for all prime moduli. This will guarantee that

$$\text{rank}\,|AA^T|_{p_i} = \text{rank}\,|AA^T|_{p_j}$$

in each case.

CHAPTER IV
Integer Solutions to Linear Equations

1 Introduction

Consider the system of linear algebraic equations $Ax = b$, where A is an $m \times n$ integer matrix and b is an m-vector with integer components. In general, the solution vector will not necessarily have integer components. However, there are many situations in which we seek an integer solution to such a system.

As an example, consider the problem of finding an optimal solution to an integer programming problem for which some or all of the unknowns are required to be integers. See Pyle and Cline [1973] or Zlobec and Ben-Israel [1970]. If we use floating-point arithmetic in computing a solution to one of these problems, we can expect the rounding errors introduced during the computation to lead us to non-integer values. In this case, it is not clear that we can merely round the results to integers for the following reasons:

   (i) Rounding is not unique. For example, we can round "up," round "down," or round to the nearest integer.
 (ii) Even if we agree on a rounding scheme, a rounded solution need not satisfy all of the given constraints.
(iii) Even if we satisfy all of the given constraints the rounded solution need not be optimal.

It seems obvious, therefore, that we need to use error-free computation rather than floating-point computation in attempting the solution of this class of problems.

134

Another application, in which we look for integer solutions to systems of linear algebraic equations with integer coefficients, is in the branch of chemical mathematics known as stoichiometry. Here we deal with weight relations determined by chemical equations and formulas. Accordingly, the balancing of equations (both redox and non-redox) is very important. Most methods for balancing such equations are based on trial and error and utilize a set of rules for determining the change in oxidation number (Andrews and Kokes [1963] and Benson [1962]), for balancing half reactions, or for balancing "gain or loss of electrons".

In this chapter we develop a deterministic method, free from trial and error, which is suitable for mechanization. We use error-free computation to solve for the (nontrivial) integer solutions of a system of homogeneous linear algebraic equations whose coefficient matrix (called the reaction matrix) does not have full rank. Since the coefficient matrix has integer elements we use a reflexive g-inverse with special integral properties in determining a solution. See Krishnamurthy [1978].

Since this method is not based on the chemical concept of oxidation-reduction, it is not recommended for classroom use in balancing chemical equations. It is useful, however, for

(i) mechanically ruling out impossible reactions when the reaction matrix has full rank.
(ii) classifying a reaction as unique (within relative proportions) when the rank of the reaction matrix is one less than its full rank, and
(iii) classifying a reaction as non-unique when the rank of the reaction matrix is at least two less than its full rank.

Before describing this algorithm we need the background material in the next section.

2 Theoretical Background

We merely summarize the material in Ben-Israel and Greville [1974], pp. 93–96, which is based on the paper by Hurt and Waid [1970]. Let $\mathbb{I}$ denote the set of integers and recall that $(\mathbb{I}, +, \cdot)$ is a ring. We introduce the following notation:

(i) $\mathbb{I}^m$ is the set of m-dimensional vectors over $\mathbb{I}$.
(ii) $\mathbb{I}^{mn}$ is the set of $m \times n$ matrices over $\mathbb{I}$.
(iii) $\mathbb{I}_r^{mn}$ is the set of $m \times n$ matrices over $\mathbb{I}$ of rank r.

2.1. **Definition** (See Marcus and Minc [1964], p. 42.). Let $A \in \mathbb{I}^{nn}$ be nonsingular. If $A^{-1} \in \mathbb{I}^{nn}$, then A is called a *unit* matrix.

2.2. **Theorem.** *If A is a unit matrix, then A is unimodular, that is, $|\det A| = 1$.*

PROOF. The proof is left as an exercise for the reader.   □

We now consider the system of linear algebraic equations

(2.3) $Ax = b,$

with $A \in \mathbb{I}^{mn}$ and $b \in \mathbb{I}^m$. In order to determine conditions under which a solution, if it exists, satisfies $x \in \mathbb{I}^n$, we introduce an algorithm for computing a reflexive g-inverse of A based on the Smith canonical form for A. The existence of the Smith canonical form is given below in Theorem 2.5.

2.4. **Definition.** $A \in \mathbb{I}^{mn}$ and $S \in \mathbb{I}^{mn}$ are *equivalent* over $\mathbb{I}$, if there exist two unit matrices $P \in \mathbb{I}^{mm}$ and $Q \in \mathbb{I}^{nn}$ such that

$$PAQ = S.$$

2.5. **Theorem** (Existence). *Let $A \in \mathbb{I}^{mn}$. Then A is equivalent over $\mathbb{I}$ to a unique matrix $S = (s_{ij}) \in \mathbb{I}_r^{mn}$ such that*

(i) $s_{ii} > 0, \quad i = 1, 2, \ldots, r$

(ii) $s_{ij} = 0, \quad otherwise,$

and s_{ii} divides $s_{i+1,i+1}$ for $i = 1, 2, \ldots, r-1$. The matrix S is called the Smith canonical form for A.

PROOF. See Marcus and Minc [1964], pp. 44–46.   □

A reflexive g-inverse of A with special integral properties can be constructed from a knowledge of the Smith canonical form for A. This is shown in the following corollaries.

2.6. **Corollary.** *Let $S = PAQ$ be the Smith canonical form for A given in Theorem 2.5. Then $G = QS^+P$ is a reflexive g-inverse of A.*

PROOF.

$$PAQ = S$$
$$= SS^+S$$
$$= (PAQ)S^+(PAQ)$$
$$= PA(QS^+P)AQ$$
$$= P(AGA)Q$$

which implies $AGA = A$. Also,

$$GAG = (QS^+P)A(QS^+P)$$
$$= QS^+(PAQ)S^+P$$

$$= Q(S^+ SS^+)P$$
$$= QS^+ P$$
$$= G$$

Hence, G is a reflexive g-inverse of A. □

2.7. **Corollary.** *The matrix* $G = QS^+ P$ *in Corollary 2.6 has the following properties:*

(i) $$AG \in \mathbb{I}^{mm}$$

(ii) $$GA \in \mathbb{I}^{nn}.$$

PROOF.

$$PAG = PA(QS^+ P)$$
$$= SS^+ P,$$

which implies

$$AG = P^{-1} SS^+ P \in \mathbb{I}^{mm},$$

likewise,

$$GAQ = (QS^+ P)AQ$$
$$= QS^+ S,$$

which implies

$$GA = QS^+ SQ^{-1} \in \mathbb{I}^{nn}.$$ □

Since $G = A_R^-$, fom Corollary 2.6, and since G has the additional integral properties described in Corollary 2.7, we will use the notation

(2.8) $$G = A_I^-.$$

to imply that A_R^- has the additional integral properties.

2.9. **Theorem.** *Let $A \in \mathbb{I}^{mn}$ and $b \in \mathbb{I}^m$. Assume that $Ax = b$ is consistent. Then $x \in \mathbb{I}^n$ if and only if*

$$A_I^- b \in \mathbb{I}^n,$$

in which case the most general integer solution is

$$x = A_I^- b + (I - A_I^- A)y,$$

where y is an arbitrary vector in $\mathbb{I}^n$.

PROOF. See Hurt and Waid [1970], and Bowman and Burdet [1974]. □

If we use this theorem, along with Theorem 3.3, Chapter 3, we have the obvious result.

2.10. Corollary. *Let $A \in \mathbb{I}^{mn}$ and $b \in \mathbb{I}^m$. Then the system of linear algebraic equations $Ax = b$ has an integer solution if and only if*

(i)
$$A_I^- b \in \mathbb{I}^n$$

and

(ii)
$$A A_I^- b = b,$$

in which case the most general integer solution is

$$x = A_I^- b + (I - A_I^- A)y,$$

where y is an arbitrary vector in $\mathbb{I}^n$.

2.11. Remark. If the system of equations is homogeneous, that is, if $b = 0$, then x becomes

$$x = (I - A_I^- A)y,$$

for an arbitrary vector $y \in \mathbb{I}^n$. Notice that if A is square and nonsingular, $A_I^- = A^{-1}$ and, in this special case, $x = 0$ is the only solution.

EXERCISES IV.2

1. Prove Theorem 2.2.

3 The Matrix Formulation of Chemical Equations

To illustrate the procedure we wish to describe, consider the problem of balancing the following chemical equation

(3.1) $a_1 \text{Al} + a_2 \text{HNO}_3 = a_3 \text{Al}(\text{NO}_3)_3 + a_4 \text{NO} + a_5 \text{H}_2\text{O}.$

Here aluminum reacts with nitric acid to produce aluminum nitrate, nitric oxide, and water. Our objective is to find $a_1, a_2, \ldots, a_5$ to satisfy the law of conservation of atoms and the law of conservation of electrons.

In this particular example we are considering only uncharged species which implies that the law of conservation of electrons is automatically satisfied when the law of conservation of atoms is satisfied. (If charged species are involved, we must treat the electrons explicitly as part of the system in order to satisfy the law of conservation of electrons.)

It is possible to describe this problem mathematically using a matrix formulation of the chemical equation. This gives us a set of homogeneous linear algebraic equations in the unknowns $a_1, a_2, \ldots, a_5$.

Suppose we have m distinct chemical elements and n distinct chemical compounds (these include both the reactants and the products) in a chemical reaction. We can form an $m \times n$ matrix (the reaction matrix) denoted by $R = (r_{ij})$ as follows. The positive integer $|r_{ij}|$ is defined as the number of atoms of type i in compound j, where $1 \leq i \leq m$ and $1 \leq j \leq n$. The integer r_{ij} is positive if compound j is a reactant and negative if compound j is a product. Thus, the reaction matrix for (3.1) can be constructed by forming the following table.

3.2. Table. The Number of Atoms.

	Al	HNO$_3$	Al(NO$_3$)$_3$	NO	H$_2$O
Al	1	0	-1	0	0
H	0	1	0	0	-2
N	0	1	-3	-1	0
O	0	3	-9	-1	-1

Notice that the chemical compounds in this table are ordered from left to right, just as they appear in (3.1), and the chemical elements are ordered from top to bottom as they appear in (3.1) reading from left to right. Using this convention,

$$\text{(3.3)} \qquad R = \begin{bmatrix} 1 & 0 & -1 & 0 & 0 \\ 0 & 1 & 0 & 0 & -2 \\ 0 & 1 & -3 & -1 & 0 \\ 0 & 3 & -9 & -1 & -1 \end{bmatrix}.$$

Consequently, if we define

$$\text{(3.4)} \qquad a = \begin{bmatrix} a_1 \\ a_2 \\ a_3 \\ a_4 \\ a_5 \end{bmatrix},$$

we obtain the coefficients $a_1, a_2, \ldots, a_n$ in (3.1) by solving the homogeneous system of linear algebraic equations

$$\text{(3.5)} \qquad Ra = 0.$$

In general, $R \in \mathbb{I}^{mn}$ with $m \neq n$. However, even if $m = n$ we expect R to be singular; otherwise, we have only the trivial solution, $a = 0$, which implies that no chemical reaction is possible in this case. Obviously, from Remark 2.11, any integer solution to (3.5) has the form

(3.6) $$a = Wy,$$

where

(3.7) $$W = I - R_I^- R,$$

and where $y \in \mathbb{I}^n$. It is clear that we need to use error-free arithmetic in these computations.

4 Solving the Homogeneous System

From (3.6) and (3.7), the matrix $W \in \mathbb{I}^{nn}$ is the matrix needed for solving the homogeneous system (3.5). The following result is of key importance.

4.1. **Theorem.** *If rank $R = r$ and rank $W = k$, then k is the nullity of R, that is,*

$$k = n - r.$$

PROOF. The proof is left as an exercise for the reader.   □

Obviously, either $k = 0$ or $k > 0$, and this motivates us to consider the following two cases.

 (i) If $r = n \leq m$, then $k = 0$.
(ii) If either $r < n \leq m$, or if $r \leq m < n$, then $k > 0$.

When R is the reaction matrix for a chemical reaction of the form (3.1), we can classify the reactions based on the value of r. Thus, we have three cases.

 (i) When $r = n$, $k = 0$ and the space of solutions to $Ra = 0$ has dimension zero. Hence, $a = 0$ is the only solution to (3.5), which implies that the chemical equation cannot be balanced and no chemical reaction is possible in this case.
(ii) When $r = n - 1$, $k = 1$ and the space of solutions to $Ra = 0$ has dimension one. This implies that there is only one linearly independent solution. Thus, the chemical reaction is unique within relative proportions of the reactants and products.
(iii) When $r \leq n - 2$, $k \geq 2$ and the space of solutions to $Ra = 0$ has dimension two or more, This implies that two or more linearly independent solutions are possible and the chemical reaction is not unique within relative proportions. Thus, we can balance the equation for different (linearly independent) relative proportions of the reactants and products.

The computation of R_I^-

From Corollary 2.6 we see that the procedure for computing R_I^- involves the computation of the Smith canonical form for R,

$$(4.2) \qquad\qquad S = PRQ,$$

followed by the computation

$$(4.3) \qquad\qquad R_I^- = QS^+ P.$$

To find the unit matrices P and Q we proceed as follows. We reduce R to its Smith canonical form by using one or more of the following three types of elementary operations.

(i) Interchange two rows (columns) of R.
(ii) Multiply a row (column) of R by a nonzero integer and add it to another row (column) of R.
(iii) Multiply a row (column) of R by minus one.

Each of these elementary row (column) operations is equivalent to multiplying R on the left (right) by an elementary matrix. The elementary matrices, which correspond to the three types of elementary row (column) operations mentioned above, have integer elements. Thus, since P and Q are products of these elementary matrices, $P \in \mathbb{I}^{mm}$ and $Q \in \mathbb{I}^{nn}$. Likewise, since the inverses of the elementary row (column) operations are of the same type, $P^{-1} \in \mathbb{I}^{mm}$ and $Q^{-1} \in \mathbb{I}^{nn}$, and this guarantees that P and Q are unit matrices. See Definition 2.1.

The two matrices P and Q can be obtained by beginning with two identity matrices of the appropriate orders and performing (on the one representing P) the same *row* operations which are performed on R, and performing (on the one representing Q) the same *column* operations which are performed on R.

The algorithm we use is a modified version of the Hermite algorithm used in Section 4, Chapter III. Instead of reducing the non-zero diagonal elements to 1 (which cannot be done, in general, using unit matrices) we use unit matrices and reduce the non-zero diagonal elements to integer values which successively divide each other (the Smith canonical form). First, we eliminate the off-diagonal elements of the first column and the first row, in that order. Next, we eliminate the off-diagonal elements of the second column and the second row, in that order. The process is continued until the matrix is in Smith canonical form.

4.4. EXAMPLE. We use R in (3.3). This is the reaction matrix for the chemical equation expressed in (3.1). We begin with R and the two identity matrices

$$\begin{bmatrix} 1 & 0 & 0 & 0 \\ 0 & 1 & 0 & 0 \\ 0 & 0 & 1 & 0 \\ 0 & 0 & 0 & 1 \end{bmatrix}, \quad \begin{bmatrix} 1 & 0 & -1 & 0 & 0 \\ 0 & 1 & 0 & 0 & -2 \\ 0 & 1 & -3 & -1 & 0 \\ 0 & 3 & -9 & -1 & -1 \end{bmatrix}, \quad \begin{bmatrix} 1 & 0 & 0 & 0 & 0 \\ 0 & 1 & 0 & 0 & 0 \\ 0 & 0 & 1 & 0 & 0 \\ 0 & 0 & 0 & 1 & 0 \\ 0 & 0 & 0 & 0 & 1 \end{bmatrix}.$$

The off-diagonal elements in the first column of R are zero so we shift our attention to the first row. There is only one non-zero off-diagonal element in the first row and this is eliminated as follows.

Add column 1 to column 3.

$$\begin{bmatrix} 1 & 0 & 0 & 0 & 0 \\ 0 & 1 & 0 & 0 & -2 \\ 0 & 1 & -3 & -1 & 0 \\ 0 & 3 & -9 & -1 & -1 \end{bmatrix}, \quad \begin{bmatrix} 1 & 0 & 1 & 0 & 0 \\ 0 & 1 & 0 & 0 & 0 \\ 0 & 0 & 1 & 0 & 0 \\ 0 & 0 & 0 & 1 & 0 \\ 0 & 0 & 0 & 0 & 1 \end{bmatrix}.$$

We now eliminate off-diagonal elements in the second column.

$$\begin{bmatrix} 1 & 0 & 0 & 0 \\ 0 & 1 & 0 & 0 \\ 0 & -1 & 1 & 0 \\ 0 & -3 & 0 & 1 \end{bmatrix}, \quad \begin{bmatrix} 1 & 0 & 0 & 0 & 0 \\ 0 & 1 & 0 & 0 & -2 \\ 0 & 0 & -3 & -1 & 2 \\ 0 & 0 & -9 & -1 & 5 \end{bmatrix}$$

(i) Add (-1) times row 2 to row 3.
(ii) Add (-3) times row 2 to row 4.

This is followed by eliminating off-diagonal elements in the second row.

Add 2 times column 2 to column 5.

$$\begin{bmatrix} 1 & 0 & 0 & 0 & 0 \\ 0 & 1 & 0 & 0 & 0 \\ 0 & 0 & -3 & -1 & 2 \\ 0 & 0 & -9 & -1 & 5 \end{bmatrix}, \quad \begin{bmatrix} 1 & 0 & 1 & 0 & 0 \\ 0 & 1 & 0 & 0 & 2 \\ 0 & 0 & 1 & 0 & 0 \\ 0 & 0 & 0 & 1 & 0 \\ 0 & 0 & 0 & 0 & 1 \end{bmatrix}.$$

This process is continued in the following manner.

$$\begin{bmatrix} 1 & 0 & 0 & 0 \\ 0 & 1 & 0 & 0 \\ 0 & 1 & -1 & 0 \\ 0 & 0 & -3 & 1 \end{bmatrix}, \quad \begin{bmatrix} 1 & 0 & 0 & 0 & 0 \\ 0 & 1 & 0 & 0 & 0 \\ 0 & 0 & 3 & 1 & -2 \\ 0 & 0 & 0 & 2 & -1 \end{bmatrix}$$

(i) Multiply row 3 by (-1).
(ii) Add 3 times row 3 to row 4.

Interchange column 3 with column 4.

$$\begin{bmatrix} 1 & 0 & 0 & 0 & 0 \\ 0 & 1 & 0 & 0 & 0 \\ 0 & 0 & 1 & 3 & -2 \\ 0 & 0 & 2 & 0 & -1 \end{bmatrix}, \quad \begin{bmatrix} 1 & 0 & 0 & 1 & 0 \\ 0 & 1 & 0 & 0 & 2 \\ 0 & 0 & 0 & 1 & 0 \\ 0 & 0 & 1 & 0 & 0 \\ 0 & 0 & 0 & 0 & 1 \end{bmatrix}$$

$$\begin{bmatrix} 1 & 0 & 0 & 0 \\ 0 & 1 & 0 & 0 \\ 0 & 1 & -1 & 0 \\ 0 & -2 & -1 & 1 \end{bmatrix}, \qquad \begin{bmatrix} 1 & 0 & 0 & 0 & 0 \\ 0 & 1 & 0 & 0 & 0 \\ 0 & 0 & 1 & 3 & -2 \\ 0 & 0 & 0 & -6 & 3 \end{bmatrix} \quad \begin{array}{l} \text{Add } (-2) \text{ times row 3} \\ \text{to row 4.} \end{array}$$

(i) Add (-3) times column 3 to column 4

(ii) Add 2 times column 3 to column 5.

$$\begin{bmatrix} 1 & 0 & 0 & 0 & 0 \\ 0 & 1 & 0 & 0 & 0 \\ 0 & 0 & 1 & 0 & 0 \\ 0 & 0 & 0 & -6 & 3 \end{bmatrix}, \qquad \begin{bmatrix} 1 & 0 & 0 & 1 & 0 \\ 0 & 1 & 0 & 0 & 2 \\ 0 & 0 & 0 & 1 & 0 \\ 0 & 0 & 1 & -3 & 2 \\ 0 & 0 & 0 & 0 & 1 \end{bmatrix}.$$

(i) Interchange column 4 with column 5.

(ii) Add 2 times column 4 to column 5.

$$\begin{bmatrix} 1 & 0 & 0 & 0 & 0 \\ 0 & 1 & 0 & 0 & 0 \\ 0 & 0 & 1 & 0 & 0 \\ 0 & 0 & 0 & 3 & 0 \end{bmatrix}, \qquad \begin{bmatrix} 1 & 0 & 0 & 0 & 1 \\ 0 & 1 & 0 & 2 & 4 \\ 0 & 0 & 0 & 0 & 1 \\ 0 & 0 & 1 & 2 & 1 \\ 0 & 0 & 0 & 1 & 2 \end{bmatrix}.$$

From these results we see that

$$P = \begin{bmatrix} 1 & 0 & 0 & 0 \\ 0 & 1 & 0 & 0 \\ 0 & 1 & -1 & 0 \\ 0 & -2 & -1 & 1 \end{bmatrix},$$

$$Q = \begin{bmatrix} 1 & 0 & 0 & 0 & 1 \\ 0 & 1 & 0 & 2 & 4 \\ 0 & 0 & 0 & 0 & 1 \\ 0 & 0 & 1 & 2 & 1 \\ 0 & 0 & 0 & 1 & 2 \end{bmatrix},$$

and (see Exercises III.2, Problem 6)

$$S^+ = \begin{bmatrix} 1 & 0 & 0 & 0 \\ 0 & 1 & 0 & 0 \\ 0 & 0 & 1 & 0 \\ 0 & 0 & 0 & \frac{1}{3} \\ 0 & 0 & 0 & 0 \end{bmatrix}.$$

Hence,

$$R_I^- = QS^+P$$

$$= \tfrac{1}{3} \begin{bmatrix} 3 & 0 & 0 & 0 \\ 0 & -1 & -2 & 2 \\ 0 & 0 & 0 & 0 \\ 0 & -1 & -5 & 2 \\ 0 & -2 & -1 & 1 \end{bmatrix},$$

which implies

$$W = I - R_I^- R$$

$$= \begin{bmatrix} 0 & 0 & 1 & 0 & 0 \\ 0 & 0 & 4 & 0 & 0 \\ 0 & 0 & 1 & 0 & 0 \\ 0 & 0 & 1 & 0 & 0 \\ 0 & 0 & 2 & 0 & 0 \end{bmatrix}.$$

From (3.6), the general integer solution to $Ra = 0$ is the vector $a = Wy$ for any $y \in \mathbb{I}^n$. Hence,

$$a = \begin{bmatrix} 0 & 0 & 1 & 0 & 0 \\ 0 & 0 & 4 & 0 & 0 \\ 0 & 0 & 1 & 0 & 0 \\ 0 & 0 & 1 & 0 & 0 \\ 0 & 0 & 2 & 0 & 0 \end{bmatrix} \begin{bmatrix} y_1 \\ y_2 \\ y_3 \\ y_4 \\ y_5 \end{bmatrix}$$

$$= y_3 \begin{bmatrix} 1 \\ 4 \\ 1 \\ 1 \\ 2 \end{bmatrix}.$$

For example, if $y_3 = 1$, we obtain the balanced chemical equation

$$Al + 4HNO_3 = Al(NO_3)_3 + NO + 2H_2O$$

4.5. Remark. This chemical reaction is unique within the relative proportions of the reactants and products indicated above. This is consistent with our theory because R obviously has rank $n - 1$, and this fact implies that the space of solutions of $Ra = 0$ has dimension one. (See Theorem 4.1 and the remarks which follow the theorem.)

4.6. Remark. From Corollary 2.7 we know that the matrix $W = I - R_I^- R$ has integer elements even though all elements of S^+ and $R_I^- = QS^+P$ may not be integers. Consequently, we need to use the error-free arithmetic of either Chapter I or Chapter II in obtaining W. We used residue arithmetic in Example 4.19, Chapter III, and finite-segment p-adic arithmetic in Example 7.5, Chapter II, and either approach can be used in Example 4.4 above.

Additional examples

The next two examples illustrate the case in which the chemical reaction cannot be balanced, and the case in which the chemical reaction can be balanced but non-unique relative proportions of reactants and products are possible.

4.7. EXAMPLE. Consider the chemical equation

$$a_1 NO_2 + a_2 HClO = a_3 HNO_3 + a_4 HCl.$$

For this equation

$$R = \begin{bmatrix} 1 & 0 & -1 & 0 \\ 2 & 1 & -3 & 0 \\ 0 & 1 & -1 & -1 \\ 0 & 1 & 0 & -1 \end{bmatrix}.$$

It is easily verified that $r = 4$, and since $m = n = 4$, R is non-singular. We can compute

$$R^{-1} = \begin{bmatrix} 1 & 0 & -1 & 1 \\ -2 & 1 & -1 & 1 \\ 0 & 0 & -1 & 1 \\ -2 & 1 & -1 & 0 \end{bmatrix},$$

which gives us

$$W = I - R^{-1}R$$
$$= 0,$$

and so $a = 0$. *This chemical reaction cannot be balanced.*

On the other hand, if H_2O is added as a reactant, we obtain the balanced equation

$$2NO_2 + HClO + H_2O = 2HNO_3 + HCl,$$

and this reaction is unique within relative proportions. See Krishnamurthy [1978].

4.8. EXAMPLE. Consider the chemical equation

$$a_1 KClO_3 + a_2 HCl = a_3 KCl + a_4 ClO_2 + a_5 Cl_2 + a_6 H_2O.$$

For this equation,

$$R = \begin{bmatrix} 1 & 0 & -1 & 0 & 0 & 0 \\ 1 & 1 & -1 & -1 & -2 & 0 \\ 3 & 0 & 0 & -2 & 0 & -1 \\ 0 & 1 & 0 & 0 & 0 & -2 \end{bmatrix},$$

$$R_I^- = \begin{bmatrix} 1 & 0 & 0 & 0 \\ -22 & 4 & 6 & -3 \\ 0 & 0 & 0 & 0 \\ 7 & -1 & -2 & 1 \\ -14 & 2 & 4 & -2 \\ -11 & 2 & 3 & -2 \end{bmatrix},$$

and

$$W = \begin{bmatrix} 0 & 0 & 1 & 0 & 0 & 0 \\ 0 & 0 & -18 & 16 & 8 & 0 \\ 0 & 0 & 1 & 0 & 0 & 0 \\ 0 & 0 & 6 & -4 & -2 & 0 \\ 0 & 0 & -12 & 10 & 5 & 0 \\ 0 & 0 & -9 & 8 & 4 & 0 \end{bmatrix},$$

Hence, since the general integer solution to $Ra = 0$ is the vector $a = Wy$, for any $y \in \mathbb{I}^n$, and since the fourth column of W is merely twice the fifth column (that is, there are only two linearly independent columns in W) the general solution can be expressed as the linear combination

$$a = k_1 \begin{bmatrix} 1 \\ -18 \\ 1 \\ 6 \\ -12 \\ -9 \end{bmatrix} + k_2 \begin{bmatrix} 0 \\ 8 \\ 0 \\ -2 \\ 5 \\ 4 \end{bmatrix}.$$

From the mathematical point of view, any integers k_1 and k_2 provide a solution to $Ra = 0$. Thus, for example, if $k_1 = 1$ and $k_2 = 3$, we obtain

$$a = \begin{bmatrix} 1 \\ 6 \\ 1 \\ 0 \\ 3 \\ 3 \end{bmatrix},$$

which gives us the legitimate chemical reaction

$$KClO_3 + 6HCl = KCl + 3Cl_2 + 3H_2O.$$

Likewise, for $k_1 = 2$ and $k_2 = 5$, we obtain

$$a = \begin{bmatrix} 2 \\ 4 \\ 2 \\ 2 \\ 1 \\ 2 \end{bmatrix},$$

which gives us the legitimate chemical reaction

$$2KClO_3 + 4HCl = 2KCl + 2ClO_2 + Cl_2 + 2H_2O.$$

We classify these reactions as non-unique since they are not simply proportional equations but are linearly independent equations. Obviously there are infinitely many choices of k_1 and k_2 which provide us with legitimate chemical reactions.

4.9. Remark. The results obtained in Example 4.8 verify our theory. It is easily seen that R has rank $n - 2$, and this fact implies that the space of solutions of $Ra = 0$ has dimension two.

4.10. Remarks. We have three comments to make about the material discussed in Sections 3 and 4.

(i) The matrix model presented here is analogous to the closed input-output Leontif static model used in studying certain aspects of economic equilibrium. See Kuenne [1963], for example.

(ii) When a chemical reaction is not unique we can use thermodynamical criteria to choose one of the legitimate solutions. (See Krishnamurthy and Adegbeyeni [1977] and Van Zeggeren and Storey [1970].) This corresponds to constrained optimization.

(iii) Because the elements of R_I^- are not necessarily continuous functions of the elements of R, the requirement that $R \in \mathbb{I}^{mn}$ must have rank less than n for a chemical reaction to be valid agrees with the known fact that even small changes in R may be accompanied by abnormally large and discontinuous changes in the chemical reaction.

EXERCISES IV.4

1. Prove Theorem 4.1.

2. In Example 4.7, verify the claim that, if H_2O is added as a reactant, the balanced chemical equation

$$2NO_2 + HClO + H_2O = 2HNO_3 + HCl$$

is the unique (within relative proportions) chemical reaction which is produced.

3. In Example 4.8 either verify the computation of R_I^- and W or, since R_I^- is not unique, compute another value of R_I^- and the corresponding value of W. In the latter case, show that the same results can be obtained.

4. Balance the equation

$$a_1 As_2 S_3 + a_2 H_2 O + a_3 HNO_3 = a_4 NO + a_5 H_3 AsO_4 + a_6 H_2 SO_4$$

and show that the chemical reaction is unique (within relative proportions).

5 Solving a Non-Homogeneous System

In this section we do not consider a specific application but consider the non-homogeneous system $Ax = b$ with $b \neq 0$.

5.1. EXAMPLE. Consider the singular matrix (Hurt and Waid [1970])

$$A = \begin{bmatrix} -9 & -8 & -5 \\ 6 & 5 & 2 \\ 3 & 2 & -1 \end{bmatrix}.$$

For this matrix, the Smith canonical form is

$$S = \begin{bmatrix} 1 & 0 & 0 \\ 0 & 3 & 0 \\ 0 & 0 & 0 \end{bmatrix},$$

and the reflexive g-inverse, with the special integral properties of Corollary 2.7, is

$$A_I^- = \tfrac{1}{3} \begin{bmatrix} 0 & 1 & 2 \\ 0 & -1 & -2 \\ 0 & 1 & -1 \end{bmatrix}.$$

Now we consider $Ax = b$ with several choices for the vector b.
1. Suppose

$$b = \begin{bmatrix} 2 \\ 1 \\ -2 \end{bmatrix}.$$

It is easily shown that

$$A A_I^- b = \begin{bmatrix} -4 \\ 1 \\ -2 \end{bmatrix},$$

and condition (ii) of Corollary 2.10 is not satisfied. Therefore, this system is inconsistent.

2. Suppose

$$b = \begin{bmatrix} 10 \\ -5 \\ 0 \end{bmatrix}.$$

It is easily shown that $AA_I^- b = b$ and so the system is consistent. However,

$$A_I^- b = \tfrac{5}{3} \begin{bmatrix} -1 \\ 1 \\ -1 \end{bmatrix},$$

and condition (i) of Corollary 2.10 is not satisfied. Hence, the system has no integer solutions even though it is consistent.

3. Suppose

$$b = \begin{bmatrix} -1 \\ 1 \\ 1 \end{bmatrix}.$$

Because $AA_I^- b = b$ and

$$A_I^- b = \begin{bmatrix} 1 \\ -1 \\ 0 \end{bmatrix},$$

we have satisfied both conditions of Corollary 2.10 and the system has integer solutions. Since the general integer solution is

$$x = A_I^- b + (I - A_I^- A)y,$$

for any $y \in \mathbb{I}^n$, we can write

$$x = \begin{bmatrix} 1 \\ -1 \\ 0 \end{bmatrix} + \begin{bmatrix} -3 & -3 & 0 \\ 4 & 4 & 0 \\ -1 & -1 & 0 \end{bmatrix} \begin{bmatrix} y_1 \\ y_2 \\ y_3 \end{bmatrix}$$

$$= \begin{bmatrix} 1 \\ -1 \\ 0 \end{bmatrix} + k \begin{bmatrix} -3 \\ 4 \\ -1 \end{bmatrix},$$

where $k = y_1 + y_2$ is an arbitrary integer. Hence,

$$\begin{cases} x_1 = 1 - 3k \\ x_2 = -1 + 4k \\ x_3 = -k \end{cases}$$

is the general solution.

EXERCISES IV.5

1. In Example 5.1 either verify the computation which produces A_I^- or, since A_I^- is not unique, produce another value of A_I^- and the corresponding value of $I - A_I^- A$, and show that the same results can be obtained.

2. Show that the system

$$\begin{bmatrix} 33 & 16 & 72 \\ -24 & -10 & -57 \\ 9 & 6 & 15 \end{bmatrix} \begin{bmatrix} x_1 \\ x_2 \\ x_3 \end{bmatrix} = \begin{bmatrix} 23 \\ -23 \\ 0 \end{bmatrix}$$

is consistent and obtain the integer solutions using the method of this section.

6 Solving Interval Linear Programming Problems

There is a class of problems known as *optimization problems* in which we seek to maximize or minimize a function of several variables, with these variables subject to certain constraints. For more than a century we have developed classical optimization techniques for solving such problems, and this has been helpful in solving problems in the physical sciences, engineering, and economics.

Since World War II, a large class of optimization problems has emerged in the field of economics and these problems are generally referred to as *programming problems*. Practical problems in government, in industry, and in military operations give rise to this type of formulation. Unfortunately, in general, classical optimization techniques have not been useful in solving programming problems and, as a consequence, new methods have had to be developed.

A typical example of a programming problem is one which involves the determination of optimal allocations of limited resources to meet certain specified objectives when certain restrictions (constraints) are imposed. If all the relations among the variables are linear, the programming problem is called a *linear programming problem*. The relations must be linear in both the constraints and the function to be optimized.

A special class of linear programming problems (usually called *interval linear programming problems*) can be stated as follows. (See Murthy [1976] and Rao and Mitra [1971], for example.)

6.1. PROBLEM. Maximize the objective function

$$\sum_{j=1}^{n} b_j x_j = b^T x,$$

subject to the interval constraints

$$c_i \le \sum_{j=1}^{n} a_{ij} x_j \le d_i \qquad i = 1, 2, \ldots, m.$$

In matrix language, we maximize $b^T x$ subject to the interval constraints

$$c \le Ax \le d,$$

where b and x are n-dimensional vectors, c and d are m-dimensional vectors and A is an $m \times n$ dimensional matrix.

An algorithm for the solution of this problem is based on the following definitions and theorems.

6.2. **Definition.** An interval linear programming problem (called an *IP*) is said to be *feasible* (also *consistent*), if the set

$$\mathbb{S} = \{x \in \mathbb{R}^n : c \le Ax \le d\}$$

is nonempty, in which case the elements of $\mathbb{S}$ are called feasible solutions of the *IP*.

6.3. **Definition.** A feasible *IP* is called *bounded* if

$$\max\{b^T x : x \in \mathbb{S}\}$$

is finite, in which case the *optimal solutions* of the *IP* are its feasible solutions x_0 which satisfy

$$b^T x_0 = \max\{b^T x : x \in \mathbb{S}\}.$$

6.4. **Lemma.** *A feasible IP is bounded if b is an element of the column space of A^T or equivalently, if*

$$b^T A^- A = b^T,$$

where A^- is any g-inverse of A.

PROOF. See Rao and Mitra [1971], p. 193. □

6.5. **Theorem.** *If an IP is feasible and bounded, and if rank $A = m$, then the class of optimal solutions of the IP is given by*

$$x = A^- e + (I - A^- A)z$$

where z is an arbitrary vector in $\mathbb{R}^n$, and $e \in \mathbb{R}^m$ is determined as follows:

$$e_i = \begin{cases} c_i, & if (b^T A^-)_i < 0 \\ d_i, & if (b^T A^-)_i > 0 \\ arbitrary\ in\ [c_i, d_i], & if (b^T A^-)_i = 0, \end{cases}$$

for $i = 1, 2, \ldots, m$.

PROOF. See Rao and Mitra [1971], p. 193.   □

For the cases where rank $A < m$, see Rao and Mitra [1971] and Ben-Israel and Greville [1974]. We now formalize an algorithm for solving Problem 6.1.

6.6. **Algorithm.** *Let $A = (a_{ij})$, $b^T x$, c, and d be the quantities defined in Problem 6.1 above.*

Step 1. Compute the matrices E and F using (4.8) and (4.9), Chapter III.
Step 2. Compute A_R^- using (4.12), Chapter III.
Step 3. Compute $b^T A_R^-$.
Step 4. Compute $b^T A_R^- A$.
Step 5. If $b^T = b^T A_R^- A$, write 'SOLN IS BOUNDED' and go to step 6; otherwise, go to step 9.
Step 6. Determine the components of e using the formulas in Theorem 6.5. If $(b^T A_R^-)_i = 0$, choose $e_i = d_i$.
Step 7. Compute $x = A_R^- e + (I - A_R^- A)z$ where z is arbitrary. (We choose $z_i = 1$ for $i = 1, 2, \ldots, n$.)
Step 8. Stop.
Step 9. Write 'SOLN IS NOT BOUNDED' and go to step 8.

6.7. EXAMPLE. Suppose we wish to maximize the objective function

$$f(x_1, x_2, x_3) = 4x_1 + 5x_2 + 10x_3$$

subject to the interval constraints

$$0 \le 2x_1 + x_2 + 2x_3 \le 140$$

$$0 \le 3x_1 + 4x_2 + 8x_3 \le 360.$$

In matrix language this is equivalent to maximizing the function $b^T x$ subject to the interval constraints $c \le Ax \le d$, where

$$x = \begin{bmatrix} x_1 \\ x_2 \\ x_3 \end{bmatrix}, \qquad b = \begin{bmatrix} 4 \\ 5 \\ 10 \end{bmatrix}, \qquad c = \begin{bmatrix} 0 \\ 0 \end{bmatrix}, \qquad d = \begin{bmatrix} 140 \\ 360 \end{bmatrix},$$

and

$$A = \begin{bmatrix} 2 & 1 & 2 \\ 3 & 4 & 8 \end{bmatrix}.$$

If we use the procedures described in (4.8), (4.9), and (4.12), Chapter III, we observe that rank $A = 2$ and

$$A_R^- = \tfrac{1}{5} \begin{bmatrix} 4 & -1 \\ -3 & 2 \\ 0 & 0 \end{bmatrix}.$$

Consequently,

$$AA_R^- = \begin{bmatrix} 1 & 0 \\ 0 & 1 \end{bmatrix},$$

$$A_R^- A = \begin{bmatrix} 1 & 0 & 0 \\ 0 & 1 & 2 \\ 0 & 0 & 0 \end{bmatrix},$$

and

$$(I - A_R^- A) = \begin{bmatrix} 0 & 0 & 0 \\ 0 & 0 & -2 \\ 0 & 0 & 1 \end{bmatrix}.$$

In addition, we find that

$$b^T A_R^- A = [4, 5, 10] \begin{bmatrix} 1 & 0 & 0 \\ 0 & 1 & 2 \\ 0 & 0 & 0 \end{bmatrix}$$

$$= [4, 5, 10]$$

$$= b^T,$$

and so, by Lemma 6.4, the solution is bounded.

Also, by direct computation

$$b^T A_R^- = \tfrac{1}{5}[1, 6],$$

which implies

$$(b^T A_R^-)_i > 0,$$

for $i = 1, 2$. Thus, since rank $A = 2$, we use Theorem 6.5 to deduce that

$$x = A_R^- d + (I - A_R^- A)z$$

where z arbitrary. If we choose

$$z = \begin{bmatrix} 1 \\ 1 \\ 1 \end{bmatrix},$$

we obtain

$$x = \begin{bmatrix} 40 \\ 58 \\ 1 \end{bmatrix}.$$

Observe that

$$Ax = \begin{bmatrix} 2 & 1 & 2 \\ 3 & 4 & 8 \end{bmatrix} \begin{bmatrix} 40 \\ 58 \\ 1 \end{bmatrix}$$

$$= \begin{bmatrix} 140 \\ 360 \end{bmatrix},$$

and

$$b^T x = 460.$$

6.8. EXAMPLE (A is nonsingular). Suppose we are given that

$$A = \begin{bmatrix}
2 & -1 & 0 & 0 & 0 & 0 & 0 & 0 \\
-1 & 2 & -1 & 0 & 0 & 0 & 0 & 0 \\
0 & -1 & 2 & -1 & 0 & 0 & 0 & 0 \\
0 & 0 & -1 & 2 & -1 & 0 & 0 & 0 \\
0 & 0 & 0 & -1 & 2 & -1 & 0 & 0 \\
0 & 0 & 0 & 0 & -1 & 2 & -1 & 0 \\
0 & 0 & 0 & 0 & 0 & -1 & 2 & -1 \\
0 & 0 & 0 & 0 & 0 & 0 & -1 & 2
\end{bmatrix},$$

and

$$b = \begin{bmatrix} -4 \\ 1 \\ -3 \\ 2 \\ -2 \\ 3 \\ -1 \\ 4 \end{bmatrix}, \quad c = \begin{bmatrix} -9 \\ -9 \\ -9 \\ -9 \\ -9 \\ -9 \\ -9 \\ -9 \end{bmatrix}, \quad d = \begin{bmatrix} 9 \\ 9 \\ 9 \\ 9 \\ 9 \\ 9 \\ 9 \\ 9 \end{bmatrix}.$$

A is a well-known nonsingular matrix whose inverse can be expressed in the closed form

$$A^{-1} = \frac{1}{n+1} B$$

where, if B is of order n,

$$b_{ij} = \begin{cases} i(n-i+1) & \text{if } i = j \\ b_{i,j-1} - i & \text{if } j > i \\ b_{ji} & \text{if } j < i. \end{cases}$$

See Gregory and Karney [1978], p. 48, for example.

    Consequently,

$$A^{-1} = \tfrac{1}{9} \begin{bmatrix} 8 & 7 & 6 & 5 & 4 & 3 & 2 & 1 \\ 7 & 14 & 12 & 10 & 8 & 6 & 4 & 2 \\ 6 & 12 & 18 & 15 & 12 & 9 & 6 & 3 \\ 5 & 10 & 15 & 20 & 16 & 12 & 8 & 4 \\ 4 & 8 & 12 & 16 & 20 & 15 & 10 & 5 \\ 3 & 6 & 9 & 12 & 15 & 18 & 12 & 6 \\ 2 & 4 & 6 & 8 & 10 & 12 & 14 & 7 \\ 1 & 2 & 3 & 4 & 5 & 6 & 7 & 8 \end{bmatrix}.$$

Since A is nonsingular, $I - A^{-1}A = 0$. Also, since $b^T A^{-1} A = b^T$, we know (from Lemma 6.4) that the solution is bounded. By direct computation,

$$b^T A^{-1} = \tfrac{1}{9}[-30, -24, -27, -3, 3, 27, 24, 30]$$

which implies

$$e^T = [-9, -9, -9, -9, 9, 9, 9, 9].$$

Therefore, from Theorem 6.4,

$$x^T = (A^{-1}e)^T$$
$$= [-16, -23, -21, -10, 10, 21, 23, 16].$$

Observe that

$$(Ax)^T = [-9, -9, -9, -9, 9, 9, 9, 9]$$

and that

$$b^T x = 168.$$

Exercises IV.6

1. Maximize the objective function

$$f(x_1, x_2, x_3) = 3x_1 - x_2 - 2x_3$$

subject to the interval constraints

$$-10 \le 2x_1 + x_2 + 2x_3 \le 10$$
$$-20 \le 3x_1 + 4x_2 + 8x_3 \le 20.$$

7 The Solution of Systems of Mixed-Integer Linear Equations

In this section we describe an algorithm for solving a system of mixed linear equations of the form

(7.1) $$Ax + By = c,$$

where A and B are matrices whose elements are rational numbers and c is a vector whose components are also rational numbers. The vector x must have components which are integers, but the components of the vector y may be rational numbers.

Observe that if $B = 0$, we have the diophantine system

$$(7.2) \qquad\qquad\qquad Ax = c,$$

and if $A = 0$, we have the standard system of linear equations

$$(7.3) \qquad\qquad\qquad By = c,$$

briefly discussed in Section 3, Chapter III.

Let B_R^- be a reflexive g-inverse of B. Then, from Theorem 3.3, Chapter III, (7.3) has a solution if and only if

$$(7.4) \qquad\qquad\qquad BB_R^- c = c,$$

in which case the most general solution is given by

$$(7.5) \qquad\qquad y = B_R^- c + (I - B_R^- B)z,$$

where z is an arbitrary vector of rational numbers. This is easily verified by premultiplication by B.

We now investigate the g-inverse approach to solving the diophantine system (7.2) which is similar to what we were discussing in Sections 2 and 5, the difference being that in this section we do not require that A have integer elements.

Hurt and Waid [1970] use a particular form of a g-inverse for this system but we will use a class of g-inverses suggested by Bowman and Burdet [1974], because this class leads to the classical forms of Smith and Hermite.

Similar to the results in Theorem 2.9 and Corollary 2.10, we state that (7.2) has a solution if and only if

$$(7.6) \qquad\qquad\qquad A_I^- c \in \mathbb{I}^n,$$

and

$$(7.7) \qquad\qquad\qquad AA_I^- c = c,$$

in which case the most general integer solution is given by

$$(7.8) \qquad\qquad x = A_I^- c + (I - A_I^- A)w$$

where $w \in \mathbb{I}^n$ is otherwise arbitrary.

Observe that (7.1) is equivalent to the two systems

$$(7.9) \qquad\qquad \begin{cases} Ax = q \\ By = c - q, \end{cases}$$

where the first is a diophantine system and the second is a standard linear system. We can combine solutions to the two systems in (7.9) and produce a solution to the mixed system (7.1).

Using the notation in (7.9), condition (7.4) becomes

$$(7.10) \qquad BB_R^-(c - q) = c - q,$$

that is, $By = c - q$ has a solution if and only if (7.10) is satisfied. We can rewrite (7.10) in the form

$$(7.11) \qquad (I - BB_R^-)q = (I - BB_R^-)c$$

and so only values of q which satisfy (7.11) are acceptable. (By definition, q is an integer combination of the columns of A.) Thus, if we replace q by Ax, the mixed system (7.1) [or the simultaneous systems (7.9)] have a solution if and only if the diophantine system

$$(7.12) \qquad [(I - BB_R^-)A]x = (I - BB_R^-)c$$

has a solution $x \in \mathbb{I}^n$.

For simplicity of notation, let

$$(7.13) \qquad (I - BB_R^-)A = D,$$

and

$$(7.14) \qquad (I - BB_R^-)c = d,$$

so that (7.12) becomes

$$(7.15) \qquad Dx = d.$$

From Theorem 2.9, $Dx = d$ has a solution $x \in \mathbb{I}^n$ if and only if

$$(7.16) \qquad D_I^- d \in \mathbb{I}^n$$

and

$$(7.17) \qquad DD_I^- d = d,$$

in which case the most general integer solution is

$$(7.18) \qquad x = D_I^- d + (I - D_I^- D)w,$$

where w is an arbitrary vector in $\mathbb{I}^n$. If we summarize these results we have the following theorem.

7.19. Theorem. *The system of mixed-integer linear equations*

$$Ax + By = c$$

has a solution if and only if

(i) $\qquad D_I^- d = [(I - BB_R^-)A]_I^-(I - BB_R^-)c \in \mathbb{I}^n$, and

(ii) $\qquad DD_I^- d = [(I - BB_R^-)A][(I - BB_R^-)A]_I^-(I - BB_R^-)c$

$$= (I - BB_R^-)c$$

$$= d,$$

in which case the most general solution is

$$x = D_I^- d + (I - D_I^- D)w$$

and

$$y = B_R^- c - B_R^- A D_I^- d - B_R^- A(I - D_I^- D)w + (I - B_R^- B)z$$

where z is an arbitrary vector of rational numbers and w is an arbitrary vector of integers.

See Bowman and Burdet [1974], from which this development was taken. Note that if $B = 0$ and A is square and nonsingular, the solution is

(7.20) $$x = A^{-1}c,$$

and if $A = 0$ and B is square and nonsingular,

(7.21) $$y = B^{-1}c.$$

The algorithm

We now indicate the main steps used to obtain the solutions x and y to the mixed-integer linear system (7.1), that is, the system $Ax + By = c$.

Step 1. Find a reflexive* g-inverse B_R^- for B. Then compute

$$D = (I - BB_R^-)A$$

and

$$d = (I - BB_R^-)c.$$

Step 2. Compute D_I^- as explained in Section 4.
Step 3. Compute $D_I^- d$. If the components are integers, go to step 4; otherwise go to step 6.
Step 4. Check $DD_I^- d = d$. If so, go to step 5; otherwise go to step 6.
Step 5. Compute

$$x = D_I^- d + (I - D_I^- D)w$$

$$y = B_R^- c - B_R^- A D_I^- d - B_R^- A(I - D_I^- D)w + (I - B_R^- B)z$$

and then go to step 7.
Step 6. Print 'THERE IS NO SOLUTION'; go to step 7.
Step 7. Stop.

7.22. EXAMPLE. Consider the mixed-integer linear system

---
*For example, see (4.12) and Theorem 2.5, Chapter III.

$$\begin{bmatrix} 3 & 0 & 2 \\ -2 & -1 & 1 \\ 1 & 2 & 0 \end{bmatrix} \begin{bmatrix} x_1 \\ x_2 \\ x_3 \end{bmatrix} + \begin{bmatrix} 1 & -1 & 5 & 4 \\ -1 & -3 & 3 & 0 \\ 2 & 4 & -2 & 2 \end{bmatrix} \begin{bmatrix} y_1 \\ y_2 \\ y_3 \\ y_4 \end{bmatrix} = \begin{bmatrix} 7 \\ 7 \\ 7 \end{bmatrix}$$

which we write in the form

$$Ax + By = c.$$

We seek a solution in which the components of x are integers and the components of y are rational numbers.

A reflexive g-inverse of B is

$$B_R^- = \begin{bmatrix} \frac{3}{4} & -\frac{1}{4} & 0 \\ -\frac{1}{4} & -\frac{1}{4} & 0 \\ 0 & 0 & 0 \\ 0 & 0 & 0 \end{bmatrix}$$

which gives us

$$B B_R^- = \begin{bmatrix} 1 & 0 & 0 \\ 0 & 1 & 0 \\ \frac{1}{2} & -\frac{3}{2} & 0 \end{bmatrix},$$

and

$$B_R^- B = \begin{bmatrix} 1 & 0 & 3 & 3 \\ 0 & 1 & -2 & -1 \\ 0 & 0 & 0 & 0 \\ 0 & 0 & 0 & 0 \end{bmatrix}.$$

Therefore,

$$I - B B_R^- = \begin{bmatrix} 0 & 0 & 0 \\ 0 & 0 & 0 \\ -\frac{1}{2} & \frac{3}{2} & 1 \end{bmatrix}$$

and

$$I - B_R^- B = \begin{bmatrix} 0 & 0 & -3 & -3 \\ 0 & 0 & 2 & 1 \\ 0 & 0 & 1 & 0 \\ 0 & 0 & 0 & 1 \end{bmatrix}.$$

Continuing, we find

$$B_{\bar{R}} A = \tfrac{1}{4} \begin{bmatrix} 11 & 1 & 5 \\ -1 & 1 & -3 \\ 0 & 0 & 0 \\ 0 & 0 & 0 \end{bmatrix}, \qquad d = \begin{bmatrix} 0 \\ 0 \\ 14 \end{bmatrix},$$

and

$$D = \begin{bmatrix} 0 & 0 & 0 \\ 0 & 0 & 0 \\ -\tfrac{7}{2} & \tfrac{1}{2} & \tfrac{1}{2} \end{bmatrix}.$$

We choose

$$P = \begin{bmatrix} 0 & 0 & 1 \\ 0 & 1 & 0 \\ 1 & 0 & 0 \end{bmatrix}, \quad \text{and} \quad Q = \begin{bmatrix} 1 & 1 & 1 \\ 0 & 7 & 0 \\ 0 & 0 & 7 \end{bmatrix}$$

to obtain

$$S = PDQ = \begin{bmatrix} -\tfrac{7}{2} & 0 & 0 \\ 0 & 0 & 0 \\ 0 & 0 & 0 \end{bmatrix}$$

which implies

$$S^{+} = \begin{bmatrix} -\tfrac{2}{7} & 0 & 0 \\ 0 & 0 & 0 \\ 0 & 0 & 0 \end{bmatrix}.$$

Therefore,

$$D_{I}^{-} = QS^{+}P = \begin{bmatrix} 0 & 0 & -\tfrac{2}{7} \\ 0 & 0 & 0 \\ 0 & 0 & 0 \end{bmatrix},$$

$$D_{I}^{-} D = \begin{bmatrix} 1 & -\tfrac{1}{7} & -\tfrac{1}{7} \\ 0 & 0 & 0 \\ 0 & 0 & 0 \end{bmatrix}, \qquad DD_{I}^{-} = \begin{bmatrix} 0 & 0 & 0 \\ 0 & 0 & 0 \\ 0 & 0 & 1 \end{bmatrix},$$

and

$$d = \begin{bmatrix} 0 & 0 & 0 \\ 0 & 0 & 0 \\ -\tfrac{1}{2} & \tfrac{3}{2} & 1 \end{bmatrix} \begin{bmatrix} 7 \\ 7 \\ 7 \end{bmatrix} = \begin{bmatrix} 0 \\ 0 \\ 14 \end{bmatrix}.$$

We observe that

$$D_I^- d = \begin{bmatrix} -4 \\ 0 \\ 0 \end{bmatrix}$$

contains integer components, and that

$$DD_I^- d = \begin{bmatrix} 0 \\ 0 \\ 14 \end{bmatrix} = d.$$

If we pick

$$z = \begin{bmatrix} 1 \\ 1 \\ 1 \\ 1 \end{bmatrix} \quad \text{and} \quad w = \begin{bmatrix} 7 \\ 7 \\ 7 \end{bmatrix},$$

then, with the matrices and vectors given above, we obtain

$$x = \begin{bmatrix} -4 \\ 0 \\ 0 \end{bmatrix} + \begin{bmatrix} 0 & \frac{1}{7} & \frac{1}{7} \\ 0 & 1 & 0 \\ 0 & 0 & 1 \end{bmatrix} \begin{bmatrix} 7 \\ 7 \\ 7 \end{bmatrix} = \begin{bmatrix} -2 \\ 7 \\ 7 \end{bmatrix},$$

and

$$y = \begin{bmatrix} \frac{7}{2} \\ -\frac{7}{2} \\ 0 \\ 0 \end{bmatrix} + \begin{bmatrix} 11 \\ -1 \\ 0 \\ 0 \end{bmatrix} - \begin{bmatrix} 16 \\ -4 \\ 0 \\ 0 \end{bmatrix} + \begin{bmatrix} -6 \\ 3 \\ 1 \\ 1 \end{bmatrix} = \begin{bmatrix} -\frac{15}{2} \\ \frac{5}{2} \\ 1 \\ 1 \end{bmatrix}.$$

By choosing different vectors z and w, we can generate other vectors x (with integer components) and y which satisfy the given mixed-integer linear system.

EXERCISES IV.7

1. Solve the mixed-integer linear system

$$\begin{bmatrix} 1 & 0 & 1 \\ -1 & 2 & -2 \\ 0 & 3 & -1 \end{bmatrix} \begin{bmatrix} x_1 \\ x_2 \\ x_3 \end{bmatrix} + \begin{bmatrix} 2 & 0 & 1 & -1 \\ 0 & -1 & 0 & 2 \\ 1 & 0 & 0 & 1 \end{bmatrix} \begin{bmatrix} y_1 \\ y_2 \\ y_3 \\ y_4 \end{bmatrix} = \begin{bmatrix} 4 \\ -1 \\ 6 \end{bmatrix}.$$

CHAPTER V

Iterative Matrix Inversion and the Iterative Solution of Linear Equations

1 Introduction

In Section 5 Chapter II, we describe how to compute the Hensel code $H(p, r, 1/\alpha)$ if we are given $H(p, r, \alpha)$, by using the fast iterative method based on Newton's method. It is well known that Newton's method for finding the reciprocal of a real number can be generalized to

(i) the computation of the inverse of a nonsingular matrix (see, for example, Stoer and Bulirsch [1980], p. 310, where this is called Schultz's method), and

(ii) the computation of the Moore–Penrose g-inverse of an arbitrary matrix (see, for example, Ben-Israel and Greville [1974], p. 300).

In this chapter we describe how the Newton–Schultz method is used to handle both (i) and (ii) for matrices with integer (or rational) elements using residue arithmetic and finite-segment p-adic arithmetic.

Some important features of this approach are:

(i) *The deterministic nature.* The initial approximation to the matrix inverse (or to a g-inverse) is chosen deterministically as the inverse (or g-inverse) of the given matrix modulo p, where p is a prime. At the ith iterative step the digits of the elements of the inverse matrix modulo p^{2^i} are generated. The iterative procedure is stopped when i is "sufficiently large"; that is, when the results are order-N Farey fractions, where

$$2N^2 + 1 \leq p^r,$$

with $2^i = r$.

(ii) *The speed of the computation*. The rational elements of the inverse matrix can be determined with a quadratic (or higher) rate of convergence.

2 The Newton–Schultz Method for the Matrix Inverse

Before we describe the Newton–Schultz method, we insert some definitions and theorems related to the inverse of an integer matrix modulo m. We are interested in the case where $m = p^r$ and p is a prime. Since these results appear in Young and Gregory [1973], pp. 853–858, no proofs are included here. We assume that A, C, and E are $n \times n$ matrices with integer elements.

2.1. Definition. If $A = (a_{ij})$, then

$$|A|_m = (|a_{ij}|_m).$$

2.2. Definition. If $|AC|_m = |CA|_m = I$, and if $|C|_m = C$, then we write $C = A^{-1}(m)$ and call C a *multiplicative inverse of A modulo m*.

2.3. Theorem. *If $A^{-1}(m)$ exists, it is unique.*

Even though $A^{-1}(m)$ is unique, when it exists, more than one matrix can have the same inverse modulo m.

2.4. Theorem. *If $|A|_m = |E|_m$ and if $A^{-1}(m)$ exists, then $E^{-1}(m)$ exists and*

$$A^{-1}(m) = E^{-1}(m).$$

2.5. Definition. A is said to be *nonsingular modulo m* if and only if both $|\det A|_m \neq 0$ and $\gcd(\det A, m) = 1$. Otherwise, A is called *singular modulo m*.

2.6. Theorem. $A^{-1}(m)$ *exists if and only if A is nonsingular modulo m*.

2.7. Theorem.

$$|\det A|_m = |\det |A|_m|_m.$$

The algorithm for A^{-1}

Let A be an $n \times n$ matrix with integer elements and let A be nonsingular modulo m. It is easy to show that if $m = p^r$, where p is a prime, then A is non-singular modulo m if and only if A is nonsingular modulo p. See Problem 1.

The first step in the Newton–Schultz algorithm is to construct $A^{-1}(p)$ by any method and to use this as the initial approximation to $C = A^{-1}(m)$. Using notation analogous to (5.32) and (5.33), Chapter II, we let

$$(2.8) \qquad\qquad B_1 = A^{-1}(p)$$

and form subsequent approximations by using the iteration

$$(2.9) \qquad\qquad B_{2k} = \big| B_{2k-1}(2I - AB_{2k-1}) \big|_{p^{2k}}$$

for $k = 1, 2, \ldots, i$, where $r = 2^i$ is large enough to accomodate the rational numbers in the inverse matrix. In other words, if $\mathbb{F}_N$ is the set of order-N Farey fractions to which the elements of the inverse belong, then i should be such that

$$(2.10) \qquad\qquad p^{2^i} \geq 2N^2 + 1.$$

Suppose we introduce the notation $H(p, r, A^{-1})$ to denote the matrix whose elements are the Hensel codes which represent the elements (order-N Farey fractions) of A^{-1}. Then, if $r = 2^i$, the matrix B_{2^i}, (whose elements are integers) corresponds to the matrix of Hensel codes $H(p, 2^i, A^{-1})$, and the correspondence between respective elements is the same as the correspondence between b_{2k} and $H(p, 2^k, 1/\alpha)$ in (5.32), Chapter II. As soon as i is large enough to satisfy (2.10), we can stop iterating and convert the matrix elements into the appropriate order-N Farey fractions using the inverse mapping of Chapters I and II.

Before we show that (2.9) is valid, we illustrate the algorithm with a numerical problem.

2.11. EXAMPLE. Let

$$A = \begin{bmatrix} 1 & -1 & 2 \\ 3 & 2 & 4 \\ 0 & 1 & -2 \end{bmatrix}$$

and choose $p = 3$. Then

$$|A|_3 = \begin{bmatrix} 1 & 2 & 2 \\ 0 & 2 & 1 \\ 0 & 1 & 1 \end{bmatrix}.$$

To compute $A^{-1}(3)$ we use Gauss–Jordan elimination in $(\mathbb{I}_3, +, \cdot)$ and obtain

$$\begin{bmatrix} 1 & 2 & 2 & 1 & 0 & 0 \\ 0 & 2 & 1 & 0 & 1 & 0 \\ 0 & 1 & 1 & 0 & 0 & 1 \end{bmatrix} \rightarrow \begin{bmatrix} 1 & 0 & 0 & 1 & 0 & 1 \\ 0 & 1 & 0 & 0 & 1 & 2 \\ 0 & 0 & 1 & 0 & 2 & 2 \end{bmatrix}.$$

Consequently,

$$B_1 = A^{-1}(3) = \begin{bmatrix} 1 & 0 & 1 \\ 0 & 1 & 2 \\ 0 & 2 & 2 \end{bmatrix}.$$

We now use (2.9) and obtain, successively,

$$B_2 = A^{-1}(9) = \begin{bmatrix} 1 & 0 & 1 \\ 6 & 7 & 2 \\ 3 & 8 & 5 \end{bmatrix},$$

$$B_4 = A^{-1}(81) = \begin{bmatrix} 1 & 0 & 1 \\ 60 & 61 & 20 \\ 30 & 71 & 50 \end{bmatrix},$$

$$B_8 = A^{-1}(6561) = \begin{bmatrix} 1 & 0 & 1 \\ 4920 & 4921 & 1640 \\ 2460 & 5741 & 4100 \end{bmatrix},$$

and

$$B_{16} = A^{-1}(43046721) = \begin{bmatrix} 1 & 0 & 1 \\ 32285040 & 32285041 & 10761680 \\ 16142520 & 37665881 & 26904200 \end{bmatrix}.$$

When we use the inverse mapping of Chapter I, we observe that both B_8 and B_{16} produce the same matrix. Consequently, we terminate the iteration at this point.

Corresponding to the matrices B_1, B_2, B_4, and B_8 we have the matrices of Hensel codes

$$H(3, 1, B_1) = \begin{bmatrix} .1 & .0 & .1 \\ .0 & .1 & .2 \\ .0 & .2 & .2 \end{bmatrix},$$

$$H(3, 2, B_2) = \begin{bmatrix} .10 & .00 & .10 \\ .02 & .12 & .20 \\ .01 & .22 & .21 \end{bmatrix},$$

$$H(3, 4, B_4) = \begin{bmatrix} .1000 & .0000 & .1000 \\ .0202 & .1202 & .2020 \\ .0101 & .2212 & .2121 \end{bmatrix},$$

and

$$H(3, 8, B_8) = \begin{bmatrix} .10000000 & .00000000 & .10000000 \\ .02020202 & .12020202 & .20202020 \\ .01010101 & .22121212 & .21212121 \end{bmatrix}.$$

To illustrate the computation using the inverse mapping, we compute the
(2, 1) element of A^{-1}. The Hensel code .02020202 corresponds to the integer

$$20202020_{\text{three}} = 4920_{\text{ten}}$$

in B_8. Hence,

$$
\begin{array}{c|cc}
 & 6561 & 0 \\
 & 4920 & 1 \\
\hline
1 & 1641 & -1 \\
2 & 1638 & 3 \\
1 & 3 & -4 \\
\hline
546 & 0 \quad 2187
\end{array}
$$

and the order-57 Farey fraction produced is $-\frac{3}{4}$. It turns out that, when all
the elements are computed, we obtain

$$
A^{-1} = \begin{bmatrix} 1 & 0 & 1 \\ -\frac{3}{4} & \frac{1}{4} & -\frac{1}{4} \\ -\frac{3}{8} & \frac{1}{8} & -\frac{5}{8} \end{bmatrix}.
$$

The following theorem guarantees the validity of the Newton–Schultz
algorithm.

2.12. Theorem. *There exists a sequence of matrices*

$$\{B_1, B_2, B_3, \ldots, B_r, \ldots\}$$

where $r = 2^i$, such that

$$\left| AB_{2k} \right|_{p2k} = I, \qquad k = 0, 1, \ldots, i, \ldots,$$

where A is an $n \times n$ matrix, nonsingular modulo p, and where

$$B_{2k} = A^{-1}(p^{2^k}).$$

PROOF. We show that the sequence $\{B_{2k}\}$ can be generated recursively, and
then prove by induction that, for $k \geq 0$,

$$\left| AB_{2k} \right|_{p2k} = I.$$

The first member of the sequence, from (2.8), is

$$B_1 = A^{-1}(p).$$

Hence, by definition,

$$\left| AB_1 \right|_p = I.$$

Recall that B_1 can be obtained by Gauss–Jordan elimination.
   Now assume (inductive hypothesis) that

$$\left| AB_{2k-1} \right|_{p2k-1} = I.$$

Then, using (2.9), we obtain

$$\left|AB_{2k}\right|_{p2k} = \left|A\left|B_{2k-1}(2I - AB_{2k-1})\right|_{p2k}\right|_{p2k}$$
$$= \left|AB_{2k-1}(2I - AB_{2k-1})\right|_{p2k}$$

and if we use the fact that our inductive hypothesis implies

$$AB_{2k-1} = I + p^{2^{k-1}}E_{k-1},$$

for some residual matrix E_{k-1}, we are able to write

$$\left|AB_{2k}\right|_{p2k} = \left|(I + p^{2^{k-1}}E_{k-1})[2I - (I + p^{2^{k-1}}E_{k-1})]\right|_{p2k}$$
$$= \left|(I + p^{2^{k-1}}E_{k-1})(I - p^{2^{k-1}}E_{k-1})\right|_{p2k}$$
$$= I.$$

Since, by construction, the theorem holds for $k = 0$, and since the assumption that the result is true for $k - 1$ implies that it is true for k, it is true for all $k \geq 0$. □

The number of iterations

Obviously,* if we carry out i iterations, where i is the smallest integer satisfying

$$(2.13) \qquad\qquad p^{2^i} > 2 \prod_{j=1}^{n} \|c_j\|_2$$

(for $\|c_j\|_2 \neq 0$), where $\|c_j\|_2$ is the Euclidean norm of the jth column of A, then all the rational elements of A^{-1} will lie in $\mathbb{F}_N$ where N is the largest integer satisfying the inequality

$$(2.14) \qquad\qquad 2N^2 + 1 \leq p^{2^i}.$$

(If $\|c_j\|_2 = 0$ for some column, that column is omitted from the computation on the right side of the inequality.) This means that when we use the inverse mapping on the integers in B_{2i}, or on the Hensel codes in $H(p, r, B_{2i})$, there will be no pseudo-overflow.

The integer i in (2.13) is usually much larger than required because the inequality (2.13) is quite conservative. Thus, i iterations could involve much superfluous computation. In practice, we can choose (by some intuitive procedure) a value of i which should be sufficiently large and then check the results using i and $i + 1$ iterations. If they produce the same rational elements for A^{-1}, then i iterations are sufficient and B_{2i}, or equivalently, $H(p, 2^i, B_{2i})$, determines A^{-1} unambiguously.

2.15. **Remark.** The successive iterations generate all the elements of the matrix $H(p, 2^k, B_{2k})$ simultaneously in p-adic digit parallel fashion. (See Example 2.11 and recall Section 5, Chapter II.)

---

*See Rao, Subramanian, and Krishnamurthy [1976], and Example 4.19, Chapter III.

Higher order convergence

The rate of convergence of the algorithm we have described is quadratic. (We observe that the number of p-adic digits in the Hensel codes in Example 2.11 doubles at each step.) It is possible to modify (2.9), in a manner completely analogous to the way (5.39) is a modification of (5.33) in Chapter II, so as to achieve higher order convergence. Thus, for example, analogous to (5.39) in Chapter II, we have

$$(2.16) \qquad B_{qk} = |B_{qk-1}[I + D_{k-1}(I + D_{k-1}(I + \ldots)\ldots)]|_{p^{qk}},$$

where there are $q - 1$ terms in the nested expression, and where

$$(2.17) \qquad D_{k-1} = |I - AB_{qk-1}|_{p^{qk}}.$$

Thus, for cubic convergence, we have

$$(2.18) \qquad \begin{aligned} B_{3k} &= |B_{3k-1}[I + D_{k-1}(I + D_{k-1})]|_{p^{3k}} \\ &= |B_{3k-1}[I + (I - AB_{3k-1})(2I - AB_{3k-1})]|_{p^{3k}}. \end{aligned}$$

Choice of the prime p

We have assumed that A is nonsingular modulo p which implies A is nonsingular modulo p^r. If during the first step (when $A^{-1}(p)$ is to be computed) we discover that $A^{-1}(p)$ does not exist, we must choose another value of p and begin anew.

An alternative procedure is to use the method of rank-one modification which is as follows. Let a and b be arbitrary n-dimensional vectors and form the rank-one matrix

$$(2.19) \qquad V = ab^T.$$

Apply the algorithm to $A + V$ instead of A and obtain A^{-1} using the formula

$$(2.20) \qquad A^{-1} = (A + V)^{-1}\left[I + \frac{1}{s}V(A + V)^{-1}\right],$$

where

$$(2.21) \qquad s = 1 - b^T(A + V)^{-1}a.$$

This formula is a variation of the well-known Sherman–Morrison formula. See, for example, Householder [1964], p. 123 or Dahlquist and Björck [1974], p. 161.

EXERCISES V.2

1. Show that if $m = p^r$, where p is a prime, then A is nonsingular modulo m if and only if A is nonsingular modulo p.

2. Verify that the elements of B_8 and B_{16} (in Example 2.11) map onto the same set of rational numbers, that is, onto the set of elements in A^{-1}.

3. Let

$$A = \begin{bmatrix} 9 & -36 & 30 \\ -36 & 192 & -180 \\ 30 & -180 & 180 \end{bmatrix}.$$

Use the Newton–Schultz method (quadratic convergence) to find A^{-1}. Hint: Use $p = 3$.

4. Find A^{-1} for the matrix in Problem 3 using cubic convergence. Hint: Use $p = 2$.

5. Invert the following matrices using both the quadratic and the cubic Newton–Schultz methods.

$$\text{(a)} \begin{bmatrix} 1 & 0 & -1 & 0 \\ 2 & 1 & -3 & 0 \\ 0 & 1 & -1 & -1 \\ 0 & 1 & 0 & -1 \end{bmatrix} \qquad \text{(b)} \begin{bmatrix} 1 & 1 & 0 & 1 \\ 1 & 1 & 1 & 0 \\ 0 & 1 & 1 & 1 \\ 1 & 0 & 1 & 1 \end{bmatrix} \qquad \text{(c)} \begin{bmatrix} 1 & 2 & 3 \\ 1 & 2 & 4 \\ 1 & 1 & 1 \end{bmatrix}$$

3 Iterative Solution of a Linear System

We now consider the iterative solution of a system of linear algebraic equations of the form

$$(3.1) \qquad\qquad Ax = b,$$

where A is an $n \times n$ nonsingular matrix, b is an n-dimensional vector (both given), and x is the unknown n-dimensional vector we call the solution.

There is no loss of generality if we assume that both A and b contain integer elements and integer components, respectively, because, if they are rational numbers, they can be converted to integers by row scaling. (We do not consider systems of equations whose coefficients are irrational numbers because such a system cannot exist inside the memory of a computer; irrational numbers are not computer representable.) Let

$$(3.2) \qquad\qquad A_1 = |A|_p$$

and

$$(3.3) \qquad\qquad b_1 = |b|_p,$$

and assume that A is nonsingular modulo p, that is, that

$$(3.4) \qquad\qquad |\det A|_p \neq 0$$

and

$$(3.5) \qquad\qquad \gcd(\det A, p) = 1.$$

Obviously, (from Theorem 2.4) A_1 is also nonsingular modulo p. The first step is to compute $A_1^{-1}(p)$ and x_1, where we define

(3.6)
$$x_k = |x|_{p^k}, \qquad k = 1, 2, \ldots,$$

by solving the system

(3.7)
$$|A_1 x_1 - b_1|_p = |Ax_1 - b|_p = 0.$$

Following this step, we carry out the iteration

(3.8)
$$x_{k+1} = |(I - A_1^{-1}A)x_k + A_1^{-1}b|_{p^{k+1}}$$

for $k = 1, 2, \ldots, i$, where, for simplicity of notation, we write A_1^{-1} rather than $A_1^{-1}(p)$. If $\mathbb{F}_N$ is the set of order-N Farey fractions to which the components of x belong, then i must be large enough so that

(3.9)
$$p^i \geq 2N^2 + 1.$$

In practice we map the (integer) components of x_i and x_{i+1} onto their rational equivalents and, if we obtain the same set of rational components, we terminate the iteration.

Before we prove the validity of (3.8) we consider a numerical example.

3.10. EXAMPLE. Consider the system of linear algebraic equations

$$\begin{bmatrix} 1 & 5 \\ 6 & 4 \end{bmatrix} \begin{bmatrix} u \\ v \end{bmatrix} = \begin{bmatrix} 7 \\ 3 \end{bmatrix}.$$

Here, if $p = 3$, we have

$$A_1 = |A|_3$$

$$= \begin{bmatrix} 1 & 2 \\ 0 & 1 \end{bmatrix},$$

and if we compute in $(\mathbb{I}_3, +, \cdot)$, we obtain (using elimination)

$$\begin{bmatrix} 1 & 2 & 1 & 0 \\ 0 & 1 & 0 & 1 \end{bmatrix} \rightarrow \begin{bmatrix} 1 & 0 & 1 & 1 \\ 0 & 1 & 0 & 1 \end{bmatrix}.$$

Thus,

$$A_1^{-1}(3) = \begin{bmatrix} 1 & 1 \\ 0 & 1 \end{bmatrix}.$$

Also,

$$b_1 = \begin{bmatrix} 1 \\ 0 \end{bmatrix}$$

which implies

$$x_1 = A_1^{-1}b_1$$

$$= \begin{bmatrix} 1 \\ 0 \end{bmatrix},$$

and we have completed the initial step.

We obtain $x_2 = |x|_{p^2}$ as follows. First we compute

$$I - A_1^{-1}A = \begin{bmatrix} 1 & 0 \\ 0 & 1 \end{bmatrix} - \begin{bmatrix} 1 & 1 \\ 0 & 1 \end{bmatrix}\begin{bmatrix} 1 & 5 \\ 6 & 4 \end{bmatrix}$$

$$= \begin{bmatrix} -6 & -9 \\ -6 & -3 \end{bmatrix},$$

and

$$A_1^{-1}b = \begin{bmatrix} 1 & 1 \\ 0 & 1 \end{bmatrix}\begin{bmatrix} 7 \\ 3 \end{bmatrix}$$

$$= \begin{bmatrix} 10 \\ 3 \end{bmatrix}.$$

Then we compute

$$x_2 = \left\| |I - A_1^{-1}A|_9 x_1 + |A_1^{-1}b|_9 \right\|_9$$

$$= \left\| \begin{bmatrix} 3 & 0 \\ 3 & 6 \end{bmatrix}\begin{bmatrix} 1 \\ 0 \end{bmatrix} + \begin{bmatrix} 1 \\ 3 \end{bmatrix} \right\|_9$$

$$= \begin{bmatrix} 4 \\ 6 \end{bmatrix},$$

$$x_3 = \left\| |I - A_1^{-1}A|_{27} x_2 + |A_1^{-1}b|_{27} \right\|_{27}$$

$$= \left\| \begin{bmatrix} 21 & 18 \\ 21 & 24 \end{bmatrix}\begin{bmatrix} 4 \\ 6 \end{bmatrix} + \begin{bmatrix} 10 \\ 3 \end{bmatrix} \right\|_{27}$$

$$= \begin{bmatrix} 13 \\ 15 \end{bmatrix},$$

and

$$x_4 = \left\| |I - A_1^{-1}A|_{81} x_3 + |A_1^{-1}b|_{81} \right\|_{81}$$

$$= \left\| \begin{bmatrix} 75 & 72 \\ 75 & 78 \end{bmatrix}\begin{bmatrix} 13 \\ 15 \end{bmatrix} + \begin{bmatrix} 10 \\ 3 \end{bmatrix} \right\|_{81}$$

$$= \begin{bmatrix} 40 \\ 42 \end{bmatrix}.$$

We can terminate at this point, because

$$\begin{array}{c|cc}
 & 27 & 0 \\
 & 13 & 1 \\
\hline
2 & 1 & -2 \\
\hline
13 & 0 & 27
\end{array}
\qquad
\begin{array}{c|cc}
 & 81 & 0 \\
 & 40 & 1 \\
\hline
2 & 1 & -2 \\
\hline
40 & 0 & 81
\end{array}$$

and

$$\begin{array}{c|cc}
 & 27 & 0 \\
 & 15 & 1 \\
\hline
1 & 12 & -1 \\
1 & 3 & 2 \\
\hline
4 & 0 & -9
\end{array}
\qquad
\begin{array}{c|cc}
 & 81 & 0 \\
 & 42 & 1 \\
\hline
1 & 39 & -1 \\
1 & 3 & 2 \\
\hline
13 & 0 & -27
\end{array}$$

Consequently, both x_3 and x_4 map onto

$$x = \begin{bmatrix} -\frac{1}{2} \\ \frac{3}{2} \end{bmatrix}.$$

We now return to the validity of (3.8).

3.11. Theorem. *Let*

$$x_{k+1} = |(I - A_1^{-1}A)x_k + A_1^{-1}b|_{p^{k+1}}.$$

Then, given that

$$|Ax_k - b|_{p^k} = 0,$$

we have

$$|Ax_{k+1} - b|_{p^{k+1}} = 0.$$

PROOF.

$$\begin{aligned}
|Ax_{k+1} - b|_{p^{k+1}} &= |A|(I - A_1^{-1}A)x_k + A_1^{-1}b|_{p^{k+1}} - b|_{p^{k+1}} \\
&= |A[(I - A_1^{-1}A)x_k + A_1^{-1}b] - b|_{p^{k+1}} \\
&= |Ax_k - AA_1^{-1}Ax_k + AA_1^{-1}b - b|_{p^{k+1}} \\
&= |(Ax_k - b) - AA_1^{-1}(Ax_k - b)|_{p^{k+1}} \\
&= |(I - AA_1^{-1})(Ax_k - b)|_{p^{k+1}}.
\end{aligned}$$

By construction, $A_1 = |A|_p$. Consequently,

$$A_1 = A + pD,$$

where D is some residual matrix. Therefore,

$$\begin{aligned}
AA_1^{-1} &= (A_1 - pD)A_1^{-1} \\
&= I - pDA_1^{-1}
\end{aligned}$$

or
$$I - AA_1^{-1} = pDA_1^{-1}.$$

Also, from the hypothesis, $|Ax_k - b|_{p^k} = 0$, we conclude that
$$Ax_k - b = p^k c$$

for some residual vector c. Hence,
$$|Ax_{k+1} - b|_{p^{k+1}} = |(pDA_1^{-1})(p^k c)|_{p^{k+1}}$$
$$= 0. \qquad \qquad \square$$

If we use the fact that, from (3.7), $|A_1 x_1 - b_1|_p = 0$ at the beginning, then Theorem 3.11 allows us to use the inductive argument that (3.8) successively generates $x_2, x_3, \ldots, x_k, \ldots$ such that

(3.12) $$\qquad\qquad\qquad |Ax_k - b|_{p^k} = 0$$

for all $k \geq 2$.

3.13. Remarks.

(i) The method above has a linear rate of convergence. In other words, the solution vector is obtained in a digit-by-digit fashion in radix-p arithmetic.

(ii) The iterative formula (3.8) involves only matrix-vector multiplication and, hence, is more economical than the direct matrix inversion procedure (2.9), which involves matrix-matrix multiplication.

(iii) In a recent paper, Dixon [1982] suggests a similar algorithm to the one described in this section.

EXERCISES V.3

1. Solve the system of linear algebraic equations
$$\begin{bmatrix} 9 & -36 & 30 \\ -36 & 192 & -180 \\ 30 & -180 & 180 \end{bmatrix} \begin{bmatrix} x \\ y \\ z \end{bmatrix} = \begin{bmatrix} 1 \\ -1 \\ 0 \end{bmatrix}$$

using the method of this section.

2. Repeat Problem 1 for the following systems:

(a) $$\begin{bmatrix} 1 & 2 & 3 \\ 1 & 2 & 4 \\ 1 & 1 & 1 \end{bmatrix} \begin{bmatrix} x \\ y \\ z \end{bmatrix} = \begin{bmatrix} 12 \\ 14 \\ 6 \end{bmatrix}$$

(b) $$\begin{bmatrix} 5 & -1 & 0 \\ -1 & 5 & -1 \\ 0 & -1 & 5 \end{bmatrix} \begin{bmatrix} x \\ y \\ z \end{bmatrix} = \begin{bmatrix} 9 \\ 4 \\ -6 \end{bmatrix}$$

$$\text{(c)} \quad \begin{bmatrix} 4 & -1 & 0 & 0 & 0 \\ -1 & 4 & -1 & 0 & 0 \\ 0 & -1 & 4 & -1 & 0 \\ 0 & 0 & -1 & 4 & -1 \\ 0 & 0 & 0 & -1 & 4 \end{bmatrix} \begin{bmatrix} x \\ y \\ z \\ u \\ v \end{bmatrix} = \begin{bmatrix} 100 \\ 200 \\ 200 \\ 200 \\ 100 \end{bmatrix}.$$

4 Iterative Computation of g-inverses

As we mentioned in the introduction to this chapter, the Newton–Schultz method can be generalized to find the Moore–Penrose g-inverse of a matrix over the real or complex field. The question now arises as to whether or not the Newton–Schultz method described in Section 2 for computing the inverse of a nonsingular matrix based on finite-segment p-adic computation can be extended to obtain the Moore–Penrose g-inverse of an $m \times n$ integer (or rational) matrix $A = (a_{ij})$. (Recall Chapter III, for definitions and notation.)

We will now demonstrate that it is possible to carry out such an extension.

Let A be an $m \times n$ matrix and let

(4.1) $$M = (AA^T)^2.$$

Then, using the notation in Chapter III,

(4.2) $$M_R^- = [(AA^T)^2]_R^-,$$

where $M_R^- = (r_{ij})$, denotes a reflexive g-inverse of M.

Let r_{ij} lie in $\mathbb{F}_N$, the set of order-N Farey fractions, where k is an integer such that

(4.3) $$p^{2^k} \geq 2N^2 + 1,$$

and p is a prime chosen for computation. If we map the elements r_{ij} (order-N Farey fractions) into $\mathbb{I}_{p^{2k}}$, using the forward mapping described in Chapter I, we can represent M_R^- as a finite-segment p-adic matrix power series. Thus,

(4.4)
$$\begin{aligned} \left| M_R^- \right|_{p^{2k}} &= B_{2^k} \\ &= B_{2^0} + (B_{2^1} - B_{2^0}) + (B_{2^2} - B_{2^1}) + \cdots + (B_{2^k} - B_{2^{k-1}}), \end{aligned}$$

where

(4.5)
$$\begin{cases} B_{2^0} = \left| M_R^- \right|_p \\ B_{2^1} = \left| M_R^- \right|_{p^2} \\ \quad \vdots \\ B_{2^k} = \left| M_R^- \right|_{p^{2k}}. \end{cases}$$

Consequently, if $\left| M_R^- \right|_p$ exists and if (4.3) is satisfied, then B_{2^k} corresponds to the finite-segment p-adic representation of M_R^- in the sense that B_{2^k}

(a matrix with integer elements) corresponds to the matrix of Hensel codes

$$(4.6) \qquad H(p, 2^k, B_{2k}) = H(p, 2^k, M_R^-),$$

where the correspondence (elementwise) is the same as the correspondence in (5.32), Chapter II, or the correspondence exhibited in Example 2.11. To realize (4.4), it must be possible

(i) to compute $\left|M_R^-\right|_p = B_1$, and for $1 < i \le k$,

(ii) to compute $\left|M_R^-\right|_{p^{2^i}} = B_{2^i}$, given $B_{2^{i-1}}$.

We easily obtain B_1 by the method described in Chapter III, and a generalization of the Newton–Schultz procedure of Section 2 allows us to compute B_{2^i} from $B_{2^{i-1}}$.

4.7. **Algorithm.**

Step 1. Construct

$$\left|M_R^-\right|_p = B_1$$

using the method in Chapter III, so that

$$B_1 = \left|B_1 M B_1\right|_p$$

and

$$M = \left|M B_1 M\right|_p.$$

Step 2. For $i = 1, 2, \ldots, k$, compute

$$B_{2^i} = \left|B_{2^{i-1}}(2I - MB_{2^{i-1}})\right|_{p^{2^i}}.$$

Step 3. At this point $M_R^- = B_{2^k}$ and we compute A^+, the Moore–Penrose g-inverse of A, using

$$A^+ = A^T M_R^- A A^T.$$

Recall (4.3), Chapter III.

4.8. Remark. We now present a proof of the validity of the algorithm. Step 1 constructs B_1 such that

$$B_1 = \left|B_1 M B_1\right|_p$$
$$= \left|M_R^-\right|_p$$

and

$$M = \left|M B_1 M\right|_p,$$

and in Step 2 the subsequent iterations satisfy

(i) $$B_{2^i} = \left|B_{2^{i-1}}(2I - MB_{2^{i-1}})\right|_{p^{2^i}}$$

for $i = 1, 2, \ldots, k$.

It is required to prove that when M_R^- exists, and (4.3) is satisfied,

$$\left|MB_{2^i}M\right|_{p^{2^i}} = \left|M\right|_{p^{2^i}}$$

and

$$\left|B_{2^i}MB_{2^i}\right|_{p^{2^i}} = \left|B_{2^i}\right|_{p^{2^i}},$$

given that

$$\left|MB_{2^{i-1}}M\right|_{p^{2^{i-1}}} = \left|M\right|_{p^{2^{i-1}}}$$

and

$$\left|B_{2^{i-1}}MB_{2^{i-1}}\right|_{p^{2^{i-1}}} = \left|B_{2^i}\right|_{p^{2^{i-1}}}$$

for $i = 1, 2, \ldots$. We prove this by induction.

For $i = 1$, we have by construction (Step 1)

$$\left|MB_1 M\right|_p = \left|M\right|_p$$

and

$$\left|B_1 MB_1\right|_p = \left|B_1\right|_p.$$

Let us assume (inductive hypothesis) that

$$\left|MB_{2^{i-1}}M\right|_{p^{2^{i-1}}} = \left|M\right|_{p^{2^{i-1}}}.$$

This implies that for some K over $\mathbb{Q}$

(ii) $$MB_{2^{i-1}}M = (I + p^{2^{i-1}}K)M.$$

(Note that M is a square matrix.) Thus, using (i), we get

$$MB_{2^i}M = \left|M[B_{2^{i-1}}(2I - MB_{2^{i-1}})]M\right|_{p^{2^i}}$$
$$= \left|2MB_{2^{i-1}}M - MB_{2^{i-1}}MB_{2^{i-1}}M\right|_{p^{2^i}}.$$

Using (ii) this simplifies to

$$\left|2(I + p^{2^{i-1}}K)M - (I + p^{2^{i-1}}K)(I + p^{2^{i-1}}K)M\right|_{p^{2^i}}$$
$$= \left|(I + p^{2^{i-1}}K)(2M - M - p^{2^{i-1}}KM)\right|_{p^{2^i}}$$
$$= \left|(I + p^{2^{i-1}}K)(I - p^{2^{i-1}}K)M\right|_{p^{2^i}}$$
$$= \left|M\right|_{p^{2^i}}.$$

Similarly, for some K_1 and K_2 over $\mathbb{Q}$, we have

(iii) $$B_{2^{i-1}}MB_{2^{i-1}} = B_{2^{i-1}}(I + p^{2^{i-1}}K_1)$$

and

(iv) $$B_{2^{i-1}}MB_{2^{i-1}} = (I + p^{2^{i-1}}K_2)B_{2^{i-1}}.$$

Thus, using (i), we get

$$B_{2^i}MB_{2^i} = \left|B_{2^{i-1}}(2I - MB_{2^{i-1}})MB_{2^{i-1}}(2I - MB_{2^{i-1}})\right|_{p^{2^i}}$$
$$= \left|(2B_{2^{i-1}} - B_{2^{i-1}}MB_{2^{i-1}})M(2B_{2^{i-1}} - B_{2^{i-1}}MB_{2^{i-1}})\right|_{p^{2^i}}.$$

Using (iii) and (iv), this reduces to

$$\left|(2B_{2i-1} - B_{2i-1} - p^{2^{i-1}}K_2 B_{2i-1})M(2B_{2i-1} - B_{2i-1} - p^{2^{i-1}}B_{2i-1}K_1)\right|_{p^{2i}}$$
$$= \left|(I - p^{2^{i-1}}K_2)B_{2i-1}MB_{2i-1}(I - p^{2^{i-1}}K_1)\right|_{p^{2i}}$$
$$= \left|(I - p^{2^{i-1}}K_2)B_{2i-1}(I + p^{2^{i-1}}K_1)(I - p^{2^{i-1}}K_1)\right|_{p^{2i}}$$

by using (iii)

$$= \left|(I - p^{2^{i-1}}K_2)B_{2i-1}\right|_{p^{2i}}$$
$$= \left|[2I - (I + p^{2^{i-1}}K_2)]B_{2i-1}\right|_{p^{2i}}$$
$$= \left|(2B_{2i-1} - B_{2i-1}MB_{2i-1})\right|_{p^{2i}}$$

by using (iv)

$$= \left|B_{2i-1}(2I - MB_{2i-1})\right|_{p^{2i}}$$
$$= \left|B_{2i}\right|_{p^{2i}},$$

by using (i).

4.9. EXAMPLE. We use the matrix in Example 4.19, Chapter III, where A is the singular matrix

$$A = \begin{bmatrix} 1 & 0 & 1 \\ 1 & 1 & 0 \\ 1 & 0 & 1 \end{bmatrix}.$$

We choose $p = 5$ and, using the method of Example 4.19, Chapter III, we obtain

$$F = \begin{bmatrix} 1 & 0 & 0 \\ 1 & 1 & 0 \\ 4 & 0 & 1 \end{bmatrix}, \qquad R = \begin{bmatrix} 1 & 0 & 0 \\ 0 & 1 & 0 \\ 0 & 0 & 0 \end{bmatrix},$$

and

$$E = \begin{bmatrix} 4 & 0 & 0 \\ 3 & 3 & 0 \\ 4 & 0 & 1 \end{bmatrix}.$$

Consequently,

$$B_1 = \left|M_R^-\right|_5 = \left|F^T RE\right|_5$$
$$= \begin{bmatrix} 2 & 3 & 0 \\ 3 & 3 & 0 \\ 0 & 0 & 0 \end{bmatrix}.$$

Using the iterative formula in Step 2, we obtain

$$B_2 = |M_R^-|_{25} = \begin{bmatrix} 17 & 8 & 0 \\ 8 & 13 & 0 \\ 0 & 0 & 0 \end{bmatrix},$$

and

$$B_4 = |M_R^-|_{625} = \begin{bmatrix} 417 & 208 & 0 \\ 208 & 313 & 0 \\ 0 & 0 & 0 \end{bmatrix}.$$

At this point we may terminate the iteration, because if we use the inverse mapping of Chapter I on the elements of B_2 and B_4, we find that in both cases we obtain the same results. For example, for the (1, 1) element, we obtain

	25	0
	17	1
1	8	−1
2	:1	3:
8	0	−25

	625	0
	417	1
1	208	−1
2	:1	3:
208	0	−625

When the mapping is performed for all the elements, we obtain, in agreement with Example 4.19, Chapter III,

$$M_R^- = \begin{bmatrix} \frac{1}{3} & -\frac{1}{3} & 0 \\ -\frac{1}{3} & \frac{1}{2} & 0 \\ 0 & 0 & 0 \end{bmatrix},$$

and

$$A^+ = \frac{1}{6} \begin{bmatrix} 1 & 2 & 1 \\ -1 & 4 & -1 \\ 2 & -2 & 2 \end{bmatrix}.$$

4.10. Remark. It is possible to solve a system of consistent linear algebraic equations $Ax = b$, where A is singular, by using the iteration formula (3.8) with A_1^{-1} replaced by

$$B_1 = |M_R^-|_p.$$

EXERCISES V.4

1. Compute a reflexive g-inverse of

$$A = \begin{bmatrix} 1 & 0 & 0 & -1 & 0 \\ 2 & 1 & 1 & -3 & 0 \\ 0 & 1 & 2 & -1 & -1 \\ 0 & 1 & 0 & 0 & -1 \end{bmatrix},$$

and

$$B = \begin{bmatrix} 2 & 0 & 0 & 0 & -1 & 0 \\ 3 & 0 & 0 & 0 & 0 & -1 \\ 0 & 2 & 1 & 0 & -3 & -2 \\ 0 & 1 & 3 & -1 & -4 & -4 \\ 0 & 0 & 1 & -1 & 0 & 0 \end{bmatrix}$$

using the method of this section.

2. Find A^+ if

$$A = \begin{bmatrix} 5 & 0 & 6 \\ 3 & 3 & 0 \\ 5 & 0 & 6 \end{bmatrix}.$$

CHAPTER VI

The Exact Computation of the Characteristic Polynomial of a Matrix

1 Introduction

It is not recommended, in general, that we compute the coefficients of the characteristic polynomial of a matrix as a first step in finding the eigenvalues of the matrix by polynomial root-finding techniques. This is due to the fact that if ordinary floating-point arithmetic is used, the accumulation of rounding errors will produce only approximations to the coefficients and, if the polynomial is ill-conditioned, the roots of the "approximate characteristic equation" may not be good approximations to the roots of the characteristic equation. See Wilkinson [1963], Chapter 2, for a discussion of the *condition* of a polynomial equation with respect to the computation of its roots.

Nevertheless, Pennington [1970], p. 397, recommends the computation of the characteristic polynomial, followed by polynomial root-finding, as a means of finding all the eigenvalues of a matrix. He recommends that the coefficients of the characteristic polynomial be computed using the Leverrier–Faddeev method (see Faddeev and Faddeeva [1963], for example) which is related to the Decell–Leverrier method mentioned in Chapter III, Section 4. Unfortunately, he does not warn his readers of the potential danger if his approach is followed. Consequently, it is strongly recommended, if the coefficients of the characteristic polynomial are computed (whether for the purpose of computing eigenvalues or for some other reason), that the computation be free of rounding errors.

There is no loss of generality if we restrict our discussion to integer matrices because if the matrix elements are rational numbers, they can be converted to integers by scaling. What we consider in this chapter is the exact computation of the coefficients of the characteristic polynomial of an integer matrix using residue arithmetic. The algorithm is described in Rao [1978].

2 The Algorithm Applied to Lower Hessenberg Matrices

Suppose we are given the matrix (in lower Hessenberg form)

$$(2.1) \qquad A = \begin{bmatrix} a_{11} & a_{12} & & & & \\ a_{21} & a_{22} & a_{23} & & & \\ a_{31} & a_{32} & a_{33} & a_{34} & & \\ \vdots & \vdots & \vdots & \vdots & \cdots & a_{n-1,n} \\ a_{n1} & a_{n2} & a_{n3} & a_{n4} & \cdots & a_{nn} \end{bmatrix}$$

where all elements on the first superdiagonal are nonzero,* and all elements above the first superdiagonal (not shown) are zero.

Let x be an eigenvector of A corresponding to an eigenvalue λ. Then $Ax = \lambda x$, or

$$(2.2) \qquad \begin{bmatrix} (a_{11} - \lambda) & a_{12} & & & & \\ a_{21} & (a_{22} - \lambda) & a_{23} & & & \\ a_{31} & a_{32} & (a_{33} - \lambda) & a_{34} & & \\ \vdots & \vdots & \vdots & \vdots & \cdots & a_{n-1,n} \\ a_{n1} & a_{n2} & a_{n3} & a_{n4} & \cdots & (a_{nn} - \lambda) \end{bmatrix} \begin{bmatrix} x_1 \\ x_2 \\ x_3 \\ \vdots \\ x_n \end{bmatrix} = \begin{bmatrix} 0 \\ 0 \\ 0 \\ \vdots \\ 0 \end{bmatrix}.$$

Without loss of generality let $x_1 = 1$. Then

$$(2.3) \qquad \begin{cases} (a_{11} - \lambda) + a_{12}x_2 & = 0 \\ a_{21} + (a_{22} - \lambda)x_2 + a_{23}x_3 & = 0 \\ a_{31} + a_{32}x_2 + (a_{33} - \lambda)x_3 + a_{34}x_4 & = 0 \\ \cdots\cdots\cdots\cdots\cdots\cdots\cdots\cdots\cdots\cdots\cdots\cdots\cdots \\ a_{n1} + a_{n2}x_2 + a_{n3}x_3 + \cdots + (a_{nn} - \lambda)x_n = 0. \end{cases}$$

If we solve the first equation for x_2, we obtain

$$(2.4) \qquad x_2 = \frac{1}{a_{12}}\lambda - \frac{a_{11}}{a_{12}},$$

and this is a polynomial in λ of *degree one*. If we substitute this expression for x_2 into the second equation and solve for x_3, we obtain

---

*If $a_{i,i+1} = 0$ for some value of i, the matrix is partitioned and we work on the diagonal sub-matrices separately.

$$(2.5) \qquad x_3 = \frac{1}{a_{12}a_{23}}\lambda^2 - \frac{a_{22} + a_{11}}{a_{12}a_{23}}\lambda + \frac{a_{11}a_{22} - a_{21}a_{12}}{a_{12}a_{23}},$$

and this is a polynomial in λ of *degree two*. If we continue this process and solve equation $n - 1$ for x_n and substitute the expressions for $x_2, x_3, \ldots, x_n$ into equation n, we obtain a polynomial in λ of *degree n* which assumes the value zero if and only if λ is an eigenvalue of A. Hence, our polynomial of degree n is a scalar multiple of the characteristic polynomial of A.

Suppose we let $P_{i+1}(\lambda)$ denote a polynomial (in λ) of degree i, and suppose we let P_{i+1} be an $(n + 1)$-dimensional row vector whose components are the coefficients of $P_{i+1}(\lambda)$. Then, if we define

$$(2.6) \qquad P_1 = [0, 0, \ldots, 0, 0, 1],$$

we can write

$$(2.7) \qquad P_2 = \left[0, 0, \ldots, 0, \frac{1}{a_{12}}, -\frac{a_{11}}{a_{12}}\right],$$

$$(2.8) \qquad P_3 = \left[0, \ldots, \frac{1}{a_{12}a_{23}}, -\frac{(a_{11} + a_{22})}{a_{12}a_{23}}, \frac{(a_{11}a_{22} - a_{21}a_{12})}{a_{12}a_{23}}\right],$$

and so on. Observe that

$$(2.9) \qquad P_2 = -\frac{a_{11}}{a_{12}}P_1 + \frac{1}{a_{12}}\tilde{P}_1$$

where $\tilde{P}_1 = [0, 0, \ldots, 0, 1, 0]$, that is, $\tilde{P}_1$ is obtained from P_1 by performing a single cyclic left shift of the components of P_1. Similarly,

$$(2.10) \qquad P_3 = -\frac{a_{21}}{a_{23}}P_1 - \frac{a_{22}}{a_{23}}P_2 + \frac{1}{a_{23}}\tilde{P}_2$$

where $\tilde{P}_2$ is obtained from P_2 by performing a single cyclic left shift of the components of P_2. In general,

$$(2.11) \qquad P_{i+1} = \sum_{j=1}^{i} \frac{-a_{ij}}{a_{i,i+1}}P_j + \frac{1}{a_{i,i+1}}\tilde{P}_i$$

for $i = 1, 2, \ldots, n - 1$, and for $i = n$,

$$(2.12) \qquad P_{n+1} = \sum_{j=1}^{n} a_{nj}P_j - \tilde{P}_n.$$

From P_{n+1} it is easy to obtain the characteristic polynomial.

2.13. EXAMPLE (Rao). Suppose we begin with the integer matrix

$$B = \begin{bmatrix} 4 & 3 & 2 \\ 2 & 3 & 4 \\ 4 & 3 & 5 \end{bmatrix}.$$

We reduce B to upper Hessenberg form using the similarity transformation

$$A = SBS^{-1}$$

$$= \begin{bmatrix} 1 & 0 & 0 \\ 0 & 1 & \frac{2}{3} \\ 0 & 0 & 1 \end{bmatrix} \begin{bmatrix} 4 & 3 & 2 \\ 2 & 3 & 4 \\ 4 & 3 & 5 \end{bmatrix} \begin{bmatrix} 1 & 0 & 0 \\ 0 & 1 & -\frac{2}{3} \\ 0 & 0 & 1 \end{bmatrix}$$

$$= \begin{bmatrix} 4 & 3 & 0 \\ \frac{14}{3} & 5 & 4 \\ 4 & 3 & 3 \end{bmatrix}.$$

In this case,

$$P_1 = [0, 0, 0, 1]$$

$$P_2 = -\tfrac{4}{3}[0, 0, 0, 1] + \tfrac{1}{3}[0, 0, 1, 0]$$

$$= [0, 0, \tfrac{1}{3}, -\tfrac{4}{3}],$$

$$P_3 = -\tfrac{7}{6}[0, 0, 0, 1] - \tfrac{5}{4}[0, 0, \tfrac{1}{3}, -\tfrac{4}{3}] + \tfrac{1}{4}[0, \tfrac{1}{3}, -\tfrac{4}{3}, 0]$$

$$= [0, \tfrac{1}{12}, -\tfrac{3}{4}, \tfrac{1}{2}],$$

and

$$P_4 = 4[0, 0, 0, 1] + 3[0, 0, \tfrac{1}{3}, -\tfrac{4}{3}] + 3[0, \tfrac{1}{12}, -\tfrac{3}{4}, \tfrac{1}{2}] - [\tfrac{1}{12}, -\tfrac{3}{4}, \tfrac{1}{2}, 0]$$

$$= [-\tfrac{1}{12}, 1, -\tfrac{7}{4}, \tfrac{3}{2}].$$

Therefore, corresponding to P_4, we have the polynomial

$$P_4(\lambda) = -\tfrac{1}{12}\lambda^3 + \lambda^2 - \tfrac{7}{4}\lambda + \tfrac{3}{2}.$$

Since B is an integer matrix, its characteristic polynomial has integer coefficients and, in particular, the coefficient of λ^3 is unity. Since B is similar to A, their characteristic polynomials differ only by a scalar multiple. Hence, the characteristic polynomial of B can be obtained from $P_4(\lambda)$ if we multiply by -12. Consequently,

$$(-12) \cdot P_4(\lambda) = \lambda^3 - 12\lambda^3 + 21\lambda - 18$$

is the characteristic polynomial for B.

2.14. Remark. If we wish to carry out this computation in the finite field $(\mathbb{I}_p, +, \cdot)$, we need a procedure for selecting a sufficiently large prime p. A very conservative inequality appears in Rao [1978]. It is recommended that p be chosen large enough to satisfy

$$p \geq 2 \max[m^n, n(n + 1)m^{n-1}],$$

where, for any matrix norm,

$$m = \max[\|AA^T\|, \operatorname{tr}(AA^T)].$$

2.15. EXAMPLE. We use the matrix B of Example 2.13 and, even though it does not satisfy the inequality in Remark 2.14, we find that we can use $p = 101$. First, we write

$$|B|_{101} = \begin{bmatrix} 4 & 3 & 2 \\ 2 & 3 & 4 \\ 4 & 3 & 5 \end{bmatrix}.$$

Then, we add 33 times column 2 to column 3 (which annihilates the element $b_{13} = 2$), and complete the similarity transformation by adding 68 times row 3 to row 2. (The additive inverse of 33 is 68.) Thus,

$$|A|_{101} = |SBS^{-1}|_{101}$$

$$= \left\| \begin{bmatrix} 1 & 0 & 0 \\ 0 & 1 & 68 \\ 0 & 0 & 1 \end{bmatrix} \begin{bmatrix} 4 & 3 & 2 \\ 2 & 3 & 4 \\ 4 & 3 & 5 \end{bmatrix} \begin{bmatrix} 1 & 0 & 0 \\ 0 & 1 & 33 \\ 0 & 0 & 1 \end{bmatrix} \right\|_{101}$$

$$= \begin{bmatrix} 4 & 3 & 0 \\ 72 & 5 & 4 \\ 4 & 3 & 3 \end{bmatrix}$$

is in lower Hessenberg form.

Beginning with $|P_1|_{101} = [0, 0, 0, 1]$, we successively compute

$$|P_2|_{101} = [0, 0, 34, 66],$$
$$|P_3|_{101} = [0, 59, 75, 51],$$

and

$$|P_4|_{101} = [42, 1, 74, 52].$$

Since, $42^{-1}(101) = 89$,

$$|89 \cdot P_4|_{101} = [1, 89, 21, 83],$$

and, since symmetric residues carry the proper algebraic signs,

$$/89 \cdot P_4/_{101} = [1, -12, 21, -18].$$

Hence,

$$\lambda^3 - 12\lambda^2 + 21\lambda - 18$$

is the characteristic polynomial for B.

2.16. **Remark.** In Section 1 we state that we can scale matrices whose elements are rational numbers and thereby convert the matrix to an integer matrix before we begin our computation. It should be pointed out that this is not necessary. If we use the techniques described in Chapter I, Section 5, we can begin with a matrix whose elements are rational numbers.

EXERCISES VI.2

1. Find the characteristic polynomial for

$$A = \begin{bmatrix} 3 & 2 & 5 \\ 6 & -5 & 3 \\ -24 & 38 & 2 \end{bmatrix}.$$

2. Find the characteristic polynomial for

$$A = \begin{bmatrix} \frac{5}{2} & 1 & \frac{1}{2} \\ \frac{1}{2} & 2 & -\frac{1}{2} \\ -\frac{1}{2} & -1 & \frac{3}{2} \end{bmatrix}.$$

Bibliography

Alparslan, E. Finite p-adic number systems with possible applications. Ph.D. Dissertation. Department of Electrical Engineering, University of Maryland, College Park, 1975.

Andrews, D. H. and Kokes, R. J. *Fundamental Chemistry*. John Wiley, New York, 1963.

Bachman, G. *Introduction to p-adic Numbers and Valuation Theory*. Academic Press, New York, 1964.

Beiser, P. S. An examination of finite-segment p-adic number systems as an alternative methodology for performing exact arithmetic. M.S. Thesis. Department of Applied Mathematics and Computer Science, University of Virginia, Charlottesville, 1979.

Ben-Israel, A. and Greville, T. N. E. *Generalized Inverses: Theory and Applications*. Wiley-Interscience, New York, 1974. Reprinted with corrections by Robert E. Kreiger Pub. Co., Melbourne, Florida, 1980.

Benson, S. W. *Chemical Calculations*. John Wiley, New York, 1962.

Bhimasankaram, P. Some contributions to the theory, application, and computation of generalized inverses of matrices. Ph.D. Dissertation. Indian Statistical Institute, Calcutta, 1971.

Boullion, T. L. and Odell, P. L. *Generalized Inverse Matrices*. Wiley-Interscience, New York, 1971.

Bowman, V. J. and Burdet, C. A. On the general solution of systems of mixed-integer linear equations. *SIAM J. Appl. Math.*, **26**, 1974, 120–125.

Dahlquist, G. and Björk, A. *Numerical Methods*. (translated by N. Anderson). Prentice-Hall, Englewood Cliffs, N.J., 1974.

Decell, H. P. Jr. An application of the Cayley–Hamilton theorem to generalized matrix inversion. *SIAM Review*, **7**, 1965, 526–528.

Dixon, J. D. Exact solution of linear equations using p-adic expansions. *Numer. Math.*, **40**, 1982, 137–141.

Faddeev, D. K. and Faddeeva, V. N. *Computational Methods of Linear Algebra*. W. H. Freeman, San Francisco, 1963.

Farinmade, J. A. Fast parallel exact matrix computations using p-adic arithmetic.

M.S. Thesis. Department of Computer Sciences, University of Lagos, Nigeria, 1976.

Forsythe, G. E., Malcom, M. A., and Moler, C. B. *Computer Methods for Mathematical Computations*. Prentice-Hall, Englewood Cliffs, N.J., 1977.

Gregory, R. T. The use of finite-segment p-adic arithmetic for exact computation. *BIT*, **18**, 1978, 282–300.

Gregory, R. T. *Error-Free Computation*. Robert E. Krieger Pub. Co., Melbourne, Florida, 1980.

Gregory, R. T. Residue arithmetic with rational operands. Proceedings 5th Symposium on Computer Arithmetic. IEEE Computer Society, Ann Arbor, Michigan, 144–145, 1981a.

Gregory, R. T. Error-free computation with rational numbers. *BIT*, **21**, 1981b, 194–202.

Gregory, R. T. A method for and an application of error-free computation. Proceedings of the AFCET Symposium "Mathematics for Computer Science." Paris, 152–158, 1982.

Gregory, R. T. Exact computation with order-N Farey fractions. *Computer Science and Statistics: Proceedings of the 15th Symposium on the Interface*. J. E. Gentle Editor. North Holland, Amsterdam, 1983.

Gregory, R. T. and Karney, D. L. *A Collection of Matrices for Testing Computational Algorithms*. Robert E. Krieger Pub. Co., Melbourne, Florida, 1978.

Greville, T. N. E. Some applications of the pseudo-inverse of a matrix, *SIAM Review*, **2**, 1960, 15–22.

Hardy, G. H. and Wright, E. M. *An Introduction to the Theory of Numbers* (4th ed.). Clarendon Press, Oxford, 1960.

Hehner, E. C. R. and Horspool, R. N. S. A new representation of the rational numbers for fast easy arithmetic. *SIAM J. Comp.*, **8**, 1979, 124–134.

Hensel, K. *Theorie der Algebraischen Zahlen*. Teubner, Leipzig, 1908.

Householder, A. S. *The Theory of Matrices in Numerical Analysis*. Blaisdell Pub. Co., New York, 1964.

Howell, J. A. On the reduction of a matrix to Frobenius form using residue arithmetic. Ph.D. Dissertation. Department of Computer Sciences, University of Texas, Austin, 1971.

Howell, J. A. and Gregory, R. T. An algorithm for solving linear algebraic equations using residue arithmetic, Parts I and II. *BIT*, **9**, 1969, 220–224 and 324–337.

Howell, J. A. and Gregory, R. T. Solving linear equations using residue arithmetic— Algorithm II. *BIT*, **10**, 1970, 23–37.

Hurt, M. F. and Waid, C. A generalized inverse which gives all the integral solutions to a system of linear equations. *SIAM J. Appl. Math.*, **19**, 1970, 547–550.

Hwang, Shu-Hwa Computation in a finite field using rational operands. M. S. Thesis. Department of Computer Science, University of Tennessee, Knoxville, 1981.

Knuth, D. E. *The Art of Computer Programming, Vol. II, Seminumerical Algorithms* (2nd ed.). Addison-Wesley, Reading, Massachusetts, 1981.

Koblitz, N. *p-adic Numbers, p-adic Analysis and Zeta-functions*. Springer–Verlag, New York, 1977.

Kornerup P. and Gregory, R. T. Mapping integers and Hensel codes onto Farey fractions. *BIT*, **23**, 1983, 9–20.

Krishnamurthy, E. V. On optimal iterative schemes for high speed division. IEEE Transactions on Computers, C-19, 1970, 227–231.

Krishnamurthy, E. V. Economical iterative and range transformation schemes for division. IEEE Transactions on Computers, C-20, 1971, 470–472.

Krishnamurthy, E. V. Matrix processors using p-adic arithmetic for exact linear computations. IEEE Transactions on Computers, C-26, 1977, 633–639.

Krishnamurthy, E. V. Generalized matrix inverse approach for automatic balancing of chemical equations. *Inter. J. Math. Educ. Sci Technol.*, **9**, 1978, 323–328.

Krishnamurthy, E. V. Fast parallel exact computation of the Moore–Penrose inverse and rank of a matrix. *Elektronische Informationsverarbeitung und Kybernetik*, **19**, 1983, 95–98.

Krishnamurthy, E. V. and Adegbeyeni, E. O. Finite field computational techniques for the exact solutions of mixed-integer linear equations. *Inter J. Syst. Sci.*, **8**, 1977, 1181–1192.

Krishnamurthy, E. V. and Klette, R. Fast parallel realization of matrix multiplication. *Electronische Informationsverarbeitung und Kybernetik*, **17**, 1981, 279–292.

Krishnamurthy, E. V. and Murthy, V. K. Fast iterative division of p-adic numbers. IEEE Transactions on Computers, C-32, 1983, 396–398.

Krishnamurthy, E. V., Rao, T. M., and Subramanian, K. Finite-segment p-adic number systems with applications to exact computation, *Proc. Indian Acad. Sci.*, 81A, 1975a, 58–79.

Krishnamurthy, E. V., Rao, T. M., and Subramanian, K. p-adic arithmetic procedures for exact matrix computations, *Proc. Indian Acad. Sci.*, 82A, 1975b, 165–175.

Krishnamurthy, E. V. and Sen, S. K. *Computer-based Numerical Algorithms*. East-west Press, New Delhi, 1976.

Kuenne, R. E. *The Theory of General Economic Equilibrium*. Princeton University Press, Princeton, N.J., 1963.

Lewis, Ruth Ann. p-adic number systems for error-free computation. Ph.D. Dissertation. Department of Mathematics, University of Tennessee, Knoxville, 1979.

MacDuffee, C. C. The p-adic numbers of Hensel, *Amer. Math. Monthly*, **45**, 1938, 500–508.

Mahler, K. *Introduction to p-adic Numbers and Their Functions*. Cambridge University Press, Cambridge, 1973.

Marcus, M. and Minc, H. *A Survey of Matrix Theory and Matrix Inequalities*. Allyn and Bacon, Boston, 1964.

McCoy, N. H. *Rings and Ideals*. Carus Monography #8. The Mathematical Association of America, Washington, D.C., 1948.

Miola, A. M. The conversion of Hensel codes to their rational equivalents. *ACM Sigsam Bulletin*, Nov. 1982, 24–26.

Murthy, K. G. *Linear and Combinatorial Programming*. John Wiley, New York, 1976.

Pennington, R. H. *Introductory Computer Methods and Numerical Analysis* (2nd ed.). Macmillan, Toronto, 1970.

Pettofrezzo, A. J. and Byrkit, D. R. *Elements of Number Theory*. Prentice-Hall, Englewood Cliffs, N.J., 1970.

Pyle, L. D. and Cline, R. E. The generalized inverse in linear programming—interior gradient projection methods, *SIAM J. Appl. Math.*, **24**, 1973, 511–534.

Rao, C. R. and Mitra, S. K. *Generalized Inverse of Matrices and its Applications*. John Wiley, New York, 1971.

Rao, T. M. Finite field computational techniques for exact solution of numerical problems. Ph.D. Dissertation. Department of Applied Mathematics, Indian Institute of Science, Bangalore, 1975.

Rao, T. M. Error-free computation of characteristic polynomial of a matrix. *Comp. and Math. with Appl.*, **4**, 1978, 61–65.

Rao, T. M. and Gregory, R. T. The conversion of Hensel codes to rational numbers. Proceedings 5th Symposium on Computer Arithmetic. IEEE Computer Society, Ann Arbor, Michigan, 1981, 10–20.

Rao, T. M., Subramanian, K., and Krishnamurthy, E. V. Residue arithmetic algorithms for exact computation of g-inverses of matrices, *SIAM J. Numer. Anal.*, **13**, 1976, 155–171.

Smyre, J. S. Exact computation using extended-precision single-modulus residue arithmetic. M.S. Thesis. Department of Computer Science, University of Tennessee, Knoxville, 1983.

Stallings, W. T. and Boullion, T. L. Computation of pseudoinverse matrices using residue arithmetic, *SIAM Review*, **14**, 1972, 152–163.

Stewart, G. W. On the continuity of the generalized inverse, *SIAM J. Appl. Math.*, **17**, 1969, 33–45.

Stoer, J. and Bulirsch, R. *An Introduction to Numerical Analysis*. Springer–Verlag, New York, 1980.

Subramanian, K. Symbolic processing of polynomial matrices using finite field transforms. Ph.D. Dissertation. School of Automation, Indian Institute of Science, Bangalore, 1977.

Szabó, N. S. and Tanaka, R. I. *Residue Arithmetic and its Applications to Computer Technology*. McGraw-Hill, New York, 1967.

Van Zeggeren, F. and Storey, S. H. *The Computation of Chemical Equilibria*. Cambridge University Press, Cambridge, 1970.

Wilkinson, J. H. *Rounding Errors in Algebraic Processes*. Prentice-Hall, Englewood Cliffs, N.J., 1963.

Young, D. M. and Gregory, R. T. *A Survey of Numerical Mathematics*, Vol. I, Addison-Wesley, Reading, Massachusetts, 1972.

Young, D. M. and Gregory, R. T. *A Survey of Numerical Mathematics*, Vol. II, Addison-Wesley, Reading, Massachusetts, 1973.

Zassenhaus, H. On Hensel Factorization, *J. of Number Theory*, **1**, 1969, 291–311.

Zlobec, S. and Ben-Israel, A. On explicit solutions of interval linear programs, *Israel J. Math.*, **8**, 1970, 12–22.

Index